Praise for *The Future of Design*

"An insightful treatment of how design must change to address the many challenges with a world of global companies and design teams."
　　—Donald Norman, author, *The Design of Everyday Things*

"This is a hugely important book that appears at a critical time."
　　—Bruce Nussbaum, Mentor-in-Residence, New Museum, NYC, former Managing Editor, *BusinessWeek*

"A must read for anyone who recognizes the need to graduate past a world of objects made of wood and metal, to design experiences that will largely be made out of computer codes."
　　—John Maeda, Global Head of Computational Design + Inclusion, Automattic, Inc.

"*The Future of Design* is an excellent overview of professional design. It's an invaluable resource for those interested in pursuing a career in the field or for entrepreneurs looking to harness the power of great design."
　　—Carole Bilson, President, Design Management Institute

"Lorraine Justice has created a guide that will help designers and those who want to learn what design is and can be. Her experience as an educator, strategist and researcher provides the base to describe the who, what and why in design."
　　—Craig M. Vogel, FIDSA, FRSA, Associate Dean, College of DAAP, University of Cincinnati

"*The Future of Design* is a timely book for designers, students in design disciplines, business, CEOs, product managers, and team players."
　　—Min Wang, Professor at Central Academy of Fine Arts, Beijing, China, founding partner at HYVC

"*The Future of Design* is an authoritative and inspiring exploration into the role of design in production and society, technology and the market."
　　—Lorenzo Imbesi, PhD, Full Professor of Design, Chair of Sapienza Design Research, Sapienza University of Rome

"An important contribution to the field of design."

> —Rama Gheerawo Director, The Helen Hamlyn Centre for Design, Royal College of Art

"Lorraine Justice provides a new interpretation of "the future of design" in this networked digital era, while the narrative structure has set up a personalized framework for design practitioners, managers, and researchers."

> —Yongqi Lou, Dean and Professor, the College of Design and Innovation, Tongji University, Executive Editor, *She Ji. The Journal of Design, Economics, and Innovation*

"Great insights with global perspective into emerging forces that affect how we design with respect to both rapid technological advancements and new meanings of lives; a very important story of design that business leaders and the general public need to understand in order to bring design into the search of creative and ethical solutions to complex issues of the contemporary world."

> —Xiangyang Xin, PhD, Professor, Pro-Rector and Dean of Graduate School, City University of Macau, Founder, XXY Innovation

THE FUTURE
OF DESIGN

Global Product Innovation
for a Complex World

By Lorraine Justice, PhD

NICHOLAS BREALEY
PUBLISHING

BOSTON • LONDON

First published in 2019 by Nicholas Brealey Publishing
An imprint of John Murray Press

An Hachette UK company

24 23 22 21 20 19   1 2 3 4 5 6 7 8 9 10

A CIP catalogue record for this title is available from the British Library

Library of Congress Cataloging-in-Publication Data
Names: Justice, Lorraine, 1955- author.
Title: The future of design : global product innovation in a complex world / by Lorraine Justice.
Description: Boston, MA : Nicholas Brealey Publishing, an imprint of John Murray Press, 2019. | Includes bibliographic references and index.
Identifiers: LCCN 2018051686 (print) | LCCN 2018060744 (ebook) | ISBN 9781473684690 (ebook) | ISBN 9781473684706 (library ebook) | ISBN 9781473684676 (pbk.) | ISBN 9781473684683 (UK ebk.)
Subjects: LCSH: Product design. | New products--Management. | Design and technology. | Globalization.
Classification: LCC TS171 (ebook) | LCC TS171 .J87 2019 (print) | DDC 658.5/752--dc23LC record available at https://lccn.loc.gov/2018051686

ISBN 978-1-4736-8467-6
US eBook 978-1-4736-8469-0
UK eBook 978-1-4736-8468-3

Printed and bound in the United States of America.

John Murray Press policy is to use papers that are natural, renewable and recyclable products and made from wood grown in sustainable forests. The logging and manufacturing processes are expected to conform to the environmental regulations of the country of origin.

John Murray Press Ltd
Carmelite House
50 Victoria Embankment
London EC4Y 0DZ
Tel: 020 3122 6000

Nicholas Brealey Publishing
Hachette Book Group
53 State Street
Boston, MA 02109, USA
Tel: (617) 263 1834

www.nbuspublishing.com

This book is dedicated to Steven, Alexa, and Ryan

Acknowledgments

T*he Future of Design* seemed natural for me to write because I work with some of the best design professionals in the world. The encouragement, insights, and expertise of my family and friends kept me on track through several drafts of this book. My literary agent, John Willig, and my acquisitions editor, Alison Hankey from Nicholas Brealey, are two of the best literary professionals in the field. They have earned my respect and friendship for being there for me at all hours and advising me to reach for the best. Alison led me to Alicia Simons and Jessica Berardi, who helped me shape my website and marketing for this book.

Alexa Justice, thank you for "tough love" on the early versions of this book, along with Steven Justice who would lovingly tell me where I was falling short on explanations. Additional thanks go to Bill Destler, Jeremy Haefner, and Jim Watters from the Rochester Institute of Technology for arranging sabbatical time off for me to write this book. They, too, are consummate professionals and academic officers.

I would like to specifically thank the following people for providing contacts, helping to shape sections of the book, and supporting me in many ways: Mary Baiardo, Tom Beckett, Jessica Bellas, Katherine Bennett, Tomas Barazza, Alex Chunn, Rick Cott, Steffi Czerny, Tim Fletcher, Brandon Gien, Elaine Gizler, Brett Halbleib, David Hanson, Christopher Keller, Mary Ann Kocher, Joseph Koncelik, Alex Lobos, James Ludwig, Heather McGowan, Patricia Moore, Michelle Morgan, Donald Norman, Bruce Nussbaum, Josh Owen, Alexander Renz, Jochen Renz, Maurizio Rossi, RitaSue Siegel, Yossi Vardi, Craig Vogel, Michele Washburn, Martin Wezowski, Robert Wolcott, Xin Xinyang, Qifeng Yan, Angela Yeh, Gino Yu.

Finally, I would like to thank those who have helped to shape the design field over the years from disciplines such as art, business, engineering, human factors, social science, and sustainability. Together, you have influenced the field of design to become a mature profession that will be ready to make significant contributions in the future.

Contents

The Future of Design

Design, at its best, *feels* good. It brings people into the moment through their senses and gives them a great experience. Whether you're designing a functional object, a temporary shelter for victims of natural disasters, a theme park, or interactive technology, well-designed products and services should make people's experiences and lives better, if not enjoyable in some way.

Good design can alleviate problematic experiences too. People want to feel safe in a hospital, hotel room, or walking down a street at night. Understanding the impact of spaces that elicit feelings of discomfort, confusion, or fear helps designers create an environment that is comfortable and reassuring. This also applies to convenience. Improving airline check-in counters, online banking, or hospital protocol can be a goal of designers when implementing the design process, an important part of which is to work directly with consumers and users on improving the experience.

Good design is efficient. It can turn cumbersome and complicated processes into clear instruction. Complex problems can be visualized through steps to help bring about solutions and opportunities. Countries such as Australia, Finland, and Sweden have included their citizens in the design process to create easier tax forms and simplify other governmental functions.

These are examples of *design thinking* in action. Design thinking is an overarching term that includes the design process and can spur large and small innovations. It encompasses two fundamental ways of thinking: *divergent* and *convergent* thinking. Divergent thinking enables you to broaden your thinking in the pursuit of ideas. Convergent thinking enables you to narrow your thoughts to eliminate solutions and ideas that do not fit the problem.

Trying to find a solution to a design problem or opportunity can be mentally taxing and intense. But it can also be exhilarating and satisfying when ideas come together. Designers and their teams need to maintain a positive yet critical mindset when considering new ideas. They synthesize large amounts of information from various sources in relation to the problem or opportunity. They actively work to connect the information to create insights that might form solutions. Using information to drive insights is a form of *design reasoning*, a key concept within the design thinking process.

Design and innovation rarely follow a linear path, so it may take several iterations of the design process to come up with something unique that addresses the original challenge. Design reasoning helps designers move between divergent thinking and convergent thinking. It is often a fast process, but it is iterative and reactive, which helps to move the designer and the team toward a solution. Design reasoning occurs when:

- Disparate pieces of information are compared for usefulness.
- Those disparate pieces of information are accepted or rejected.
- Diverse ideas are held for equal consideration.

Design reasoning is a mental give-and-take that allows you to move through many solutions in order to find the best ones. Through the design reasoning process, ideas may be evaluated and rejected, rearranged, changed, stripped of attributes, and held as contradictory solutions to the same problem if there are many ways to approach it.

Design reasoning may use visualization or sketching techniques to help determine the best shapes or attributes for the solution. Design drawings or quick sketches are used to capture and record these ideas so they can be evaluated by others. In the future, new technology will assist this reasoning process through the use of generative software that will offer many solutions and dimensions. Generative software can provide designers with several versions of shapes, colors, and textures quicker than sketching, although drawing by hand often helps designers think more creatively.

Verbal generative software is in various stages of refinement today, and it may be an integral source of information to help us choose names, explain innovations, or write copy for marketing.

As the world becomes more complex through the rise of new technologies, you may see more varied fields of design. New design skills will be needed to design for all of the senses—not just touch, aural, or visual—and for emerging experiences that will become available from the technologies. Designers may become experts in areas that will arise from these new technologies, such as virtual reality and AI. You may find yourself working on products that interact with each other yet were designed in different countries or cultures. Sensitivity to and expert knowledge of cultures will become major areas of expertise for designers.

Enriched problem-solving methods require diversity in thinking and backgrounds. Having larger teams made up of various experts and differing cultures can add new dimensions to the design and innovation conversations. Such teams add the depth and breadth of information needed for future complex design problems to mitigate risk.

The rising complexity of the world means more data is needed to help find solutions to problems and aid with decision-making. Artificial intelligence programs are already searching through and sorting massive amounts of information for the legal and health sectors. As data gathering becomes more refined, the data service expert (computer or human) may become part of the design team.

So, what might this mean for the future of design? Design problem-solving processes and practices might expand with more diverse team members and the necessary data to help solve complex problems. Technologies such as artificial intelligence (smart machine-assisted processes), and big data (massive amounts of organized information) will hopefully be honed in order to address the new problems confronting the world.

The future of design will intersect at important points along a continuum of art, science, technology, and human life. As technology becomes more embedded in our day-to-day lives (and, in our bodies and those of our pets!), design thinking and design reasoning will play a pivotal role in finding solutions to complex problems. It is more important than ever that designers embrace their increasingly complicated role in the creation of products, services, and experiences.

Design, Technology, and Humanity

Humans have survived and progressed through millennia with "designed" technology, whether it was the ancient arrowhead, prehistoric farming tools, or early city designs. Our lives have been inextricably entwined with technology and design. The hope is that design (design thinking, design reasoning, and the design process) and emerging technologies will help with today's complex problems. But designers now and in the future must engage with the new and emerging technologies in a more integrated way in order to take advantage of the opportunities that these technologies can bring.

Issues such as immigrant and refugee placement suggest increased complexity around cultural diversity and globalization in our future. Shifting values between new and old generations, changing expectations and understandings of sex and gender roles, and much longer lifespans in some parts of the world are only part of the evolving story of being a modern human. The hope is that applied designs can help to ease some of the painful personal and world events occurring today and provide more joy, comfort, beauty, and ease into our daily lives. New technologies will bring incremental change (and possibly major disruption) to every sector of product and service design. There is a great need for designers who understand this and appreciate the potential for interactions with people and their cultural differences. We must be ready to design for rapid changes and the confounding problems of our time.

Augmented reality (AR), artificial intelligence (AI), virtual reality (VR), and robotics will take us into the next era of products, services, and the future of work, which promises relief from boring, dangerous, repetitive, or unsavory tasks. Futurists, or individuals who systematically search for trends that may extend into the future, believe that these technologies will enable us to focus more on creative endeavors and complex problem solving, versus menial tasks, which will inevitably lead to a higher quality of work.

As an example, consider Sophia (see Figure 1.1). Sophia is a highly expressive robot that can carry on sophisticated conversations. Sophia was created at Hanson Robotics by Dr. David Hanson, founder and CEO, and Ben Goertzel, chief scientist.[1] Hanson Robotics creates amazingly expressive and lifelike robots that build trusted and

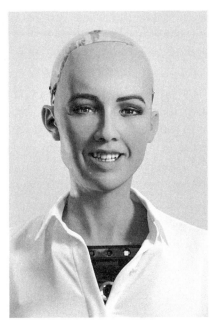

FIGURE 1.1   Sophia the lifelike robot can be viewed in action at hansonrobotics.com.

engaging relationships with people through conversation. According to Hanson's vision:

> "Our robots teach, serve, entertain and will, in time, come to truly understand and care about humans. We aim to create a better future for humanity by infusing artificial intelligence with kindness and empathy, cultivated through meaningful interactions between our robots and the individuals whose lives they touch."[2]

Sophia is a sought-after speaker and has participated as a panel member and presenter in high-level conferences, discussing how robots and artificial intelligence will become part of our lives in the future. Sophia was awarded citizenship in Saudi Arabia as a show of support for technology.

Technology will affect every area of our life, from work to leisure. As technology moves forward, it will be critical that designers and decision-makers have strong ethics in place. Consider ethical issues raised by automated driving, space travel and settlement, for example, or enhancing the human body with technologies that allow super performance. This push and pull between optimistic technologists and the "new Luddites" will no doubt inspire thoughtful discussion that will hopefully temper potentially harmful applications.

This technological future is already unfolding, in many cases, but the applications will continue to be refined and become more robust. Even doctors, lawyers, and administrators may see their work change through automation, data mining and assisted decision-making. Designers may see the creative aspects of their work shift with generative computing, where the computer will quickly generate many solutions. Designers may become arbiters of good taste and common sense, and their role may evolve into something more like a composer or film director. Computers are much better than humans at storing and comparing large amounts of data, but as technologies take over more jobs, the design of these technologies needs to be human-driven, not just based on what the technology does best.

Heather McGowan, cofounder of Work to Learn, made this observation about technology and the expanded role that automation will take in the future:

> "Designers will need to design for trust so that the individual has confidence in the automation experience. As designers moved from designing products to services and then to experiences, they must now move to designing trust in the experience . . . as things are done for you via automation."[3]

An example of designing for trust is the banking industry. Although some do not entirely trust online banking systems, they are widely used. We trust that the bank system will deposit our money in the right account, or pay a bill on time through online payments. To put it another way, designers must understand how to build trust within the automation process. But trust in technology, and in those creating the technology, might be difficult to build because many of us have had bad experiences with technology, especially where privacy and data mining are concerned.

The Future of Work

A paper by the World Economic Forum and the Boston Consulting Group titled "8 Futures of Work: Scenarios and Their Implications" calls for students to "embrace lifelong learning" as a mindset shift.[4] The paper cites McKinsey's latest report on the impact of automation and work, which "predicts between 75–375 million workers will need to change jobs or adapt their occupations by 2030."[5]

Change IS coming to the workplace, but companies, employees, and automation experts don't know how much change will occur or when a particular change will arrive. This uncertainty is exemplified in an *MIT Technology Review* article, "Every study we could find on what automation will do to jobs, in one chart," which lists "predicted job losses (and some gains), at the hands of automation, robots, and AI."[6] The author's meta-analysis concluded that there is actually no consensus and the various predictions "were across one industry or a result of a particular technology."[7] Some studies were contained to a particular product, such as the autonomous car. Other studies were lacking in depth and breadth of data. Workers might be comforted to know that disruptive technologies can take a long time to emerge. In a *South China Morning Post* article, "The hyper vision of almost every disruptive technology," Howard Yu, reminds us that it took Sony three decades before the company was able reap the benefits of transistors in radios and TVs.

> "But whether a technology is disruptive or not depends in large part on how the application is deployed, and many groundbreaking technologies are, in fact, incremental. This understanding may take a lot of excitement away from the world, but a clear vision is clearly better than a hypervision."[8]

This is all the more reason for designers and other experts to get involved in designing for technology now. Thinking about these issues today also affords us time to plan for retraining and redirecting the workforce of the future.

In January 2018, when the snow was flying in Munich, I attended the Digital Life Design (DLD) conference,[9] founded by Steffi Czerny and chaired by Yossi Vardi. While there, I spoke with Christopher Keller, head of Group Business Development of the *Telegraph*, about the future of information and technology. He spoke about "unbundling of the bundle."

In the past, people had one source, such as a newspaper, for their daily information. He said:

> "People access content differently now. In the old print media landscape, they chose what title to read based on the mix of content it offered across a broad range of topics from news, sports, and financial markets to lifestyle, cooking, and cars. The publication with the best 'bundle' of content won. In the digital world, people get their information from a much greater number of sources. They go to one website for their business news and a different website for their sports news. They go to whomever they think has the best content in a very narrowly defined topic or space. This kind of 'unbundling,' or fragmentation, means that we now have a greater number of media competing across a greater number of platforms, channels, and consumer behaviors, with different ways of consuming, inquiring, and making purchases than ever before."[10]

The shift to digital and growing market fragmentation has put a huge strain on news publishers' traditional business model, which relies heavily on advertising revenue. Effectively, it is forcing news businesses to reinvent themselves. Keller cited major newspapers such as *The Telegraph* and *The New York Times* that are expanding their offerings through mission and purpose. They are not just content providers; they have a purpose today to help their readers make better choices in life. Some are promoting curated travel tours, lectures, cooking classes, and other types of experiences. Ultimately, this means their offerings will compete with those of nonpublishing companies, such as travel agencies, educational service providers, and other businesses operating in adjacent spaces. Suddenly, companies that never competed with one another are going head-to-head. As Keller stated, "We are now thinking much more broadly about who our competitors are. Fragmentation in the digital world has meant that everyone is competing against everyone."[11]

At that same DLD conference, I spoke about the future of work with Martin Wezowski, chief designer and futurist of SAP, a German software solutions company. He expects technology:

> "to automate simplicity and augment complexity. The more mundane, boring, repetitive tasks, we can automate. However, to understand complex tasks, look for correlations, or find solutions, you want to bring in a computer or use a machine learning system. A classic

example of augmentation is using Google maps. It is a very complex thing to carry in your brain. In the middle is flow, for those cognitive tasks such as creativity, emotional intelligence, and cognitive flexibility. I call it AQ, or Adaptability Quotient, for those skills that will be much needed in the future."[12]

Wezowski surmised how machine intelligence can aid our future lifestyle.

> "Maybe we will have a digital twin that is our best buddy who will tell us we drink too much coffee in the afternoons on Tuesdays, or your Friday budget negotiations always go down by 2.7%, like a friend telling you that was the last drink of the night."[13]

Designers and other employees in the future will need to have Wezowski's adaptability in their thinking and reasoning—especially if they are going to design personalities for the new digital assistants who will be advising us or correcting our activities.

Designing Personal Digital Assistants

In 1990, my husband and I named our daughter Alexa, never imagining that 25 years later her name would be used for a digital assistant program. When visitors to my daughter's office are directed to Alexa for certain information, they are told that she is a "real person" they can interact with in the office, not a digital assistant. I recently met a woman named Siri and wondered how her life had been impacted by her technological doppelgänger! So, what are the digital assistants of today used for? The activity automation app If This Then That (ITTT) polled some of its customers about device use. The poll indicated that a majority of the personal digital assistants today are used for asking about the weather and controlling the lighting, thermostat, music, and security systems for homes.[14] More than 60 percent of the people already owned at least one connected device, such as a Fitbit.[15]

Journalist Breena Kerr set up an experiment with digital assistants from companies such as Fin, Magic, Fiverr, and Upwork. She published her results in a *New York Times* article, "The New Generation of Virtual Personal Assistants," concluding that she "wanted a personal assistant to do my work, but I didn't necessarily anticipate the work of managing a personal assistant."[16] Yes, you still need to enter information into the app and direct the work, which can seem like extra effort or make you feel as though you are halfway through the task itself. Many people continue to do the tasks themselves, but, as digital assistant technology becomes more refined and offers more, well, assistance, more consumers may demand it.

But technology poses risk as well. The threat of technology addiction, obsession, and anxiety is real. Andrew Ross Sorkin, featured in a tech article by the *New York Times*, is a self-professed "gadget junkie" who loves technology. While he probably cannot imagine life without technology, he still wants "a bigger, longer-lasting battery on the iPhone . . . and I have battery anxiety looking at the percentage icon dwindle throughout the day if I haven't found a place to plug in."[17]

Wanting a more powerful battery and checking your phone constantly might seem minor in the context of humans and technology, but Donald Norman, a pioneer in the interactive design field, believes differently. In his IASDR conference keynote "21st Century Design: A New, Unique Discipline," he highlights some of the problems:

> "The powerful human behaviors such as curiosity, attention to changes, and mental processes such as mind-wandering have been turned against us through the powers of technology. Social media, news alerts, advertisers all continually bombard us with new, fascinating tidbits. Well within the attention space, just a few words, but enough to take us on a mental journey, taking us away from more important things (such as looking for oncoming traffic while crossing a street)."[18]

You assume that technology will make your life better, but you also need to be assured that you will have a say in how that technology is designed and deployed. Technological warfare, addictive software, stolen data—what Norman provocatively calls the "evils of technology" that are difficult to stop—are here now and will continue in our future.[19] Designers, working with experts in other disciplines, need to solve for these fears, and more importantly work with experts to help stop potentially disastrous technology applications.

In his book *How to Fix the Future: Staying Human in the Digital Age*, Andrew Keen points out that "like the food industry in 1850, today's digital economy is characterized by an uncontrolled free market, addictive products, corporate irresponsibility, and wide-scale ignorance of the impact of this technology on our mental and physical well-being." He believes that citizen activism, consumer choice, responsible regulation, innovation, and education can be a few of the actions that can alleviate more problems with the uncontrolled digital market.[20]

I cannot help but agree with Keen as we see technology advancing in areas where legislation has not kept pace, and governments are underinformed about the potential outcomes. Since these technologies are not restricted to a single country, but available to the entire world, it has become difficult to know what advances have been made, especially if they are not shared or discussed.

New Interactions with Technology

What does the progress of technology mean for future designers and those working on new products and services? Deeper immersion in aspects of touch, sight, smell, voice, or combinations of these. In the past, design focus has been on the visual interface with text, images, and sound. Visual interfaces may be used more frequently for unlocking programs or spaces with facial, fingerprint, or retinal identification. New search software is capable of recognizing guns, various body postures, and threatening gestures. These products are already in limited use as research and selective application continues.

Voice recognition technologies have drastically improved during the past 20 years. While first attempts at voice recognition produced incomprehensible sentences, the new voice recognition recognizes words and also who is speaking (and it may collect data while the user interacts with products).

What about leveraging other senses for interactions? Will food be printed for that elusive flavor? Will people be able to hear and decipher languages in real time? These experiences are already here, but they are still in need of refinement before they become mainstream.

Designers will need to understand the latest technology in order to effectively design for it. It will be up to designers to help humanize technology by engaging in more user testing and creating more complex scenarios to represent our complex societies. In the words of McGowan, "We need to trust the technology we are using."[21]

Design and the Human Experience

We react emotionally to our environment. How it is designed can affect how you react. Driving a car, playing a game, or making a new recipe can enliven your senses and be a frustrating or enjoyable experience. A driver cuts you off on the highway and you feel angry; you have a terrible hand of cards and suddenly you feel bored; the recipe was not clear and now you wasted money, time, and effort. Reimagine these scenarios for a different feeling—maybe you are in a car and feel elated at seeing an incredible landscape; maybe you keep winning that card game and feel smart and smug; the new recipe you tried tasted incredible and you feel proud that your friends are admiring your cooking.

Our experience often determines how we feel about that environment, product, or service, whether it is a used toy, a piece of clothing, a visit to a nature park, or a family heirloom. Feelings about technology can differ widely and cause endless debate. For example, if you use a VR (virtual reality) product that allows you to "fly," you may feel exhilaration as you skim treetops and dive into a valley, or you may find the experience physiologically disorienting and not go for the goggles again.

Consider our experience with aging and illness, which can be diverse. A previously unhappy hospital or nursing home experience prompts many senior citizens to stay in their homes as long as possible. Patricia Moore, president of MooreDesign Associates, believes:

> "Global aging, coupled with inadequate numbers of physicians, nurses, and family care support has created a perfect storm that must be a carefully navigated blend of technology and user-centered design. Healthcare is rapidly evolving into a challenge of 'homecare,' necessitating a holistic balance of hard-tech solutions with soft-tech systems. This reality is presenting designers with the opportunity to provide for lifespan autonomy and independence, with innovations for consumers of all ages and abilities."[22]

Strides made by designers for universal aging-related products in the past were significant and include universal design "wins" such as lever door handles, ramps with texture and handrails, kitchen utensils for arthritic hands, and portable seating products for the bath. The real win was for all of society, not just the people who benefited immediately and directly from those innovations.

But how much of our human physiology will be affected by technology? Exoskeletal suits now exist that enable people to walk again, although the suits need refinement and wider distribution. For mental health-related issues such as depression, the focus has been largely on pharmaceutical offerings. Today, there are brain implants that improve memory and "work like a pacemaker, sending electrical pulses to aid the brain when it is struggling to store new information, but remain quiet when it senses that the brain is functioning well." The hope for this new technology is that it will be safe and useful—and possibly be applied to treat anxiety and depression.[23]

The Dark Future of Design

As technology continues to intersect with human activity, there is also the fear that humanity can become a victim of technologies that are "out of control," as depicted in some darkly-imagined popular science fiction stories. The futures dramatized in *Blade Runner*, *The Matrix*, or *I, Robot* are not exactly ones that many would embrace—well, maybe for the short duration of the movie, but, in reality, it is impossible to know what would happen if machines became so smart they began to act on their own, or if/when a country developed killer robots and drone warfare. It will likely require a multifaceted effort of many to avoid doomsday scenarios. Designers and other disciplines related to technology will need to take a prominent role, with values and ethics as a key part of their education. Designers and design teams in companies are often gatekeepers for appropriate practices and procedures related to the products and services they design, sometimes to the dismay of their employers. In addition, consumers want to know more than which products and services take advantage of workers or harm the environment; they want to know all the issues related to the safety and conditions of the workers. This will make for a much more complex design process, including more extensive documentation.

The Future of the Design Field

In the future, design thinking and reasoning, or the design mindset, will become a part of work and everyday life. For example, Jeremy Haefner, provost and executive vice chancellor of the University of Denver, sees new university applicants needing:

> "the fundamentals of design thinking as a foundation to their future education. Faced with a rapidly changing employment landscape due to automation, machine learning, and robotics, new graduates will need skills that are expressly human—emotional intelligence, effective communication, and ethical leadership. Already employers are asking for graduates that have mastered team-based problem solving, empathetic inquiry, and a constructive yet respectful critique process. Universities must respond to these needs, and design thinking is a curricular platform to deliver these skills."[24]

He believes these offerings are not limited to degree programs in graphic or industrial design. Rather, he sees that design thinking can be a pedagogy across the university curriculum. The design mindset is good for making correlations and substitutions while imagining what does not yet exist.[25]

Designers have a method of synthesizing large amounts of disparate information. This information can come from expert sources, visual reviews, or user testing. In the future, much of the information may come from big data. Artificial intelligence and learning systems may come up with viable solutions. These solutions will be evaluated and eventually validated through design and consumer research, and ultimately through sales and customer/user satisfaction.

The field of design is just getting started. There will be more design specialties coming as the interfaces between humans and machines, or machines and machines, expand into visual, verbal, gestural, and sensory commands. Graphic designers, interface designers, interior designers, and product designers, for instance, will be able to choose an area of virtual reality to focus on, thus becoming more specialized in those particular skills.

As Haefner suggests, primary and secondary schools are starting to integrate design projects into their curriculum. Secondary schools are offering students design experiences and have started to integrate design thinking into school projects. In Puget Sound, Washington, six high schools took part in a challenge to build a social robot to help teens with stress. The robot's purpose "is to be a companion, provide support, or gather information from people and their environments."[26] The real-life experiment of social robots is already being carried out in China, where people can tell their digital friend their deepest desires, fears, and concerns, and get a soothing answer in return.[27]

Design is everywhere. From the moment you wake up until you go to sleep for the night, your senses have interacted with hundreds, if not thousands, of objects, spaces, and surfaces. If only a few of those interactions are problematic, it can have a negative effect on your day. If your alarm doesn't go off, the electric toothbrush runs out of power, your coffee pot breaks, and the car will not start, it can seem like a pretty bad morning.

On the other hand, technology can make your life so much easier or safer. Cannot easily find your way in the airport? The design of the signs may be inadequate. Maybe one solution would be expanded airline apps that visually and verbally direct you to your gate, telling you where you can grab a particular snack along the way. The app might also tell you if you need to slow down or speed up your pace. New programs for assessing the terrain for flash floods can also alert people to get to safer areas. If such a new predictive program works, it could save the lives of people living in dangerous areas.

Our cities, our mobility options, our lifestyle technologies, and workplaces could all benefit from the design process discussed above. Tim Fletcher, president of One BusinessDesign, uses his design background to help companies visualize how mergers can occur.[28] Using various tools and exercises, he helps participants visualize the things that need to be done, both before the merger and one year into the merger. One of the participants of Fletcher's work reported: "We have been talking about the merger for a while and I have seen spreadsheets, but the merger did not seem real. After these three days, I can see the merger and feel comfortable that we can do it."[29]

Designers can help to determine the future by helping to design that future. Design is a visual and mental method of solution generation that can be applied to much more than products. Designers often work in the world of "what doesn't yet exist," and have a comfort level in this world that can benefit many people worldwide.

How to Use This Book

This book was written for those who are currently, and in the near future, designing products and services for a global marketplace. Global product development and design have become complex. This book, while not exhaustive, is meant to provide new ways of thinking about products and services in the future. There is so much more coming in the way of consumer demands. You may see rapid changes to the design profession as technology advances, but much of this technology will be used to enhance or augment the designer and the design team. Designers, manufacturers, and any company engaged in global products and services all need to recognize the complexity that is coming to the design profession. This book will explore the pertinent issues.

Through the years, the discipline of design has borrowed content and terms from the fields of art, architecture, business, engineering, futurism, psychology, and sociology. You may find that some of the terms used in this book are similar to other fields but have a slightly different meaning in design. As an example, *design research* is an activity that may follow the scientific guidelines of traditional research, but may not seem as rigorous because the method has been modified to acquire immediate feedback on a product or service.

This book is not meant to cover all definitions of design, design thinking and reasoning, or the design process. There are other books that get into detailed discussions about those topics. There are many areas of expertise in design: animation, graphic, environmental, industrial, interior, product, user experience, and more. All of these areas of expertise will be affected by the increase in new technologies and the complexity of peoples' lives.

Throughout this book, the word "product" is used when you could easily substitute service, place, or experience. It can become cumbersome to include all of these options, so product is often used as a general term. I often use the word "designer" when "design team" can be inferred. Terminology used in the design process emerged originally from other disciplines and practices such as art, business, computing, and research. For purposes of this book, I have provided the following brief definitions of how certain terms will be used:

- *Art* is considered a self-expressive activity as opposed to designing for a client.
- *Design* is considered a process of problem solving (or opportunity finding) for others and is not considered a self-expressive activity in most cases (unless the designer is creating a self-expressive product for use by others).
- A *client* employs the designer or design team. The client can be an individual or a company requesting a new design or service.
- A *customer* purchases a product but might not be the end user.
- A *user* is the person (or thing) that employs the product or experiences the service.

I kept quotes in their entirety from people I interviewed because I want the reader to hear their words. I felt this was much more powerful than rephrasing or condensing their comments.

For designers, this book is not meant to be exhaustive in explaining the design process and all of its variants. You will be familiar with much of the content of this book and how it has been condensed for a more open discussion about the future of designing products and services. My hope is that it will get you thinking about the future and how you can play a part in designing the technological tools that we will need for future work.

For futurists and artificial intelligence experts, you will know the specifics of what technology is discussed in this book, but it will hopefully bring you closer to the field of design and help you engage more designers in your projects.

Chapters 1–8 are more practical in their application for those who want to design and distribute a product or service globally. Although it has become more complicated to launch a global product, some of the practices in this book will help reduce your risk.

The Future of Design will help all professionals who are working to design, manufacture, and distribute global products or services.

The following chapters will help you to understand how the increase in complex consumer demands affects global sales and how the risks to organizations can be lowered with the right information. Finally, this book will serve as a guide for those working with designers in the future and provide insights about what to expect during that process. Designers and design teams will still develop unique, beautiful, and useful solutions to problems, but the complexity of those problems will increase.

CHAPTER 1

What Will Make a Global Product Successful?

It is not just about sales anymore. While there was a time when revenue was the only important measure of a successful global product, brands today have new corporate missions to do good things for the world, and in many ways their products are used to fulfill this promise. In this chapter, I will discuss the criteria for five of the oldest and most prestigious international product design competitions as an indicator of excellence and what makes a global product successful.

Many companies today have created or rewritten their mission statements to include their positive role in society and commerce. For example, Patagonia, a company that makes sports clothing and outdoor gear, has a mission with a philosophy that is integrated throughout the workplace: "Build the best product, cause no unnecessary harm, use business to inspire and implement solutions to the environmental crisis."[1] Another example is IKEA, a Swedish company that designs and sells furniture, kitchen appliances, and home accessories. IKEA's mission is "to create a better everyday life for the many people."[2] They do this in part by keeping their prices low so people can afford their products.

Of course, revenue is still essential, but companies are building their brand value through positive social change that consumers appreciate. This trend of positive social change will increase in the future because it is an effective way for companies to build brand loyalty. Company ethics, environmental protection, and other cultural and social issues that affect local and global markets bring great visibility through public relations opportunities as well. Positive actions by companies help you feel good about buying something you want—or at least alleviate some of the guilt from making a purchase you might not need, knowing that someone else may benefit from the purchase.

It's About the Experience

In their paper "Framework of Product Experience," Paul Hekkert and Pieter Desmet, professors and pioneers in the area of design and emotion, tell us that user and customer experience is shaped by:

- The characteristics of the user (e.g., personality, skills, background, cultural values, and motives)
- Those of the product (e.g., shape, texture, color, and behavior)
- The context (e.g., physical, social, economical in which the interaction takes place)[3]

Often a response to a product or service is "natural," but much of your reaction may be because of our cultural, societal, or even religious values, the product itself, and the context. Hekkert and Desmet break it down so we can see that many aspects go into the user experience. The user, the product, and the context are all laden with beliefs that may sway the experience to satisfactory or unsatisfactory.

Brands rely on repeated satisfactory experiences over time to build brand loyalty. If your experiences with the same product fluctuate from good to bad, you will look for another product experience. Many people feel the same way about a product with embedded technology. The initial experience with the technology may be unsatisfactory or frustrating, so they are not willing to have that experience again.

One of the most significant changes of the past decade, and one that will increase, has been the creation of products and services that can be connected to the internet of things (IoT). These products (software, appliances, electronic devices, etc.) have increased the variety of both hardware and software, the paths of distribution, and the number of international markets in which the technology is sold. Such connectivity is likely to accelerate in the future but the question will be, "Is this useful, or just more superfluous technological application we don't really need?" Companies in the future will need to explain why they used a particular technology because savvier consumers will want to know.

What are some of the top global brands?

In some ways, product categories established thousands of years ago have not changed. Tea, fruit, and spices (food and beverages); cotton, silk, and fur (clothing); gold and beads (accessories); and blue and white china (housewares) were some of the earliest-traded goods. Today, food and beverages (McDonald's and Coca-Cola), clothing and accessories (Zara and Cartier), and decor-related items (IKEA) are all reflected within global companies with brands of significant worth.

In 2018, *Forbes* magazine published a list of the world's 100 most valuable brands that had a presence in the United States.[4] According to this list, the top 10 brands in order of value are Apple, Google, Microsoft, Facebook, Amazon, Coca-Cola, Samsung, Disney, Toyota, and AT&T. Other countries such Sweden, the UK, Italy, Japan, South Korea, and China are on the list with major brands (the US presence requirement meant China's major brands Alibaba and Tencent were not on this list). Of these top 10, six are classified as technology companies, one is beverage, one is telecommunications, and one is automotive.[5]

As we will see in the next section, the top 10 brands in China, except for one alcoholic beverage called Moutai, are all heavily technology-based.

Top Chinese Brands

In his 2004 state address, Chinese Premier Hu Jintao spoke about improving the Chinese economy through increased activity in design and innovation, with the intention of creating top Chinese brands.[6] This focus instigated activities related to design and innovation in every first- and second-tier city in China. New design programs in universities, innovation parks, design centers, and art enclaves began to emerge, signaling a new era for design and innovation, and serving as a base for brand creation.

By 2017, according to a study by Kantar Millward Brown for Brandz, the most valuable brands in China were:

1. Tencent (online sales), home of WeChat (social software technology)
2. Alibaba (online retail)

3. China Mobile (telecom)
4. ICBC (financial services)
5. Baidu (technology)
6. Huawei (technology)
7. China Construction Bank (financial services)
8. Ping An (insurance services)
9. Moutai (alcohol)
10. Agricultural Bank of China (financial services)[7]

The Chinese government understands that to increase the value of their national brands, they need to increase the quality of their product and service design. All major cities in China now have design firms and companies, such as China Bridge in Shanghai, which excel at design and consumer research in China. Because of the increase in product and service design companies, and an increase in user research and attention to branding, Chinese companies such as Haier (white goods), Lenovo (computers), and Xiaomi (phones) have increased the attractiveness and usability of their products and have all received global design awards.

Tracking Global Creative Health through Brands

In other parts of Asia, India (Tata automotive), Japan (Sony), and South Korea (LG) have been home to top global brands for decades. Other Asian countries such as the Philippines, Malaysia, and Singapore have popular brands in the food and beverage category.[8]

Some of the most recognized brands (not rated on revenue) in Africa were in the category of telecommunications (MTN Group, South Africa; Econet Wireless, Zimbabwe; and Globacom, Nigeria). One of the most notable brands in the apparel industry is Anbessa from Ethiopia.[9]

In Latin America, Corona (beverage) from Mexico is a strong brand. In Brazil, the cosmetics company called natura was mentioned because of its strong commitment to society and the environment. In Columbia, the beer brand Aguila was listed.[10]

Although major exports in countries such as Russia and Australia come largely from mineral resources, they each have a vital design community, and we can expect to see more product and service brands from them in the future. Russia is making great strides in design for equipment and machinery, and Australia in retail and service brands.

The rise of successful brands in other countries is a good indication of what I consider global creative health that manifests itself in economic wealth. When one country pulls ahead in their design of new products and innovations, they can pull millions out of poverty, as the people of China have experienced.

Differences in Global Product Demand

A global product may be a success (financially and culturally) on one continent but of little note on another. Two Chinese companies considered exporting a rice cooker or a yam cooker, but the demand for yam cookers had not (and has not) yet reached significant global demand. Yam cookers are used in parts of China and in a few markets in South America, but a health food craze involving yams may have to roll around to have yam cookers in full global demand. So, rice cookers, yes, but yam cookers, not so much.

The utilitarian rice cooker is being sold successfully in many countries where rice is a staple. In 2017, 6.9 million rice cookers were sold in the United States, up from 3.1 million in 2010.[11] This appliance is a success with cooks around the world who don't want to make rice the conventional way, spending long hours at the stove. The rice cooker is simple to use, styled appropriately, and creates jobs in manufacturing.

Do high sales, ease of use, smart styling, and cultural acceptance make the rice cooker a great product? Not yet. By today's rapidly evolving standards, the rice cooker, or the companies making the rice cooker, would need to demonstrate:

- Sustainability (using less fuel for cooking; recyclable parts; environmental safety)
- Social consciousness (giving back to communities; starting health education programs)
- Sound business ethics (workers are treated well and paid appropriately; fairness in their business operations).

There is more to becoming a great global product today than sales and product functionality, and nowhere is this reflected more than in the changing criteria for global design competitions. The next section contains a review of several global design competitions and what they consider the best attributes of product and service design.

The New Criteria for Global Design Competitions

Aesthetics and quality of manufacturing used to be the hallmarks of an exceptional product. These aspects are still required and are now considered basic, but the criteria have expanded. Usability, sustainability, and corporate social responsibility have become priorities for products, which can represent an entire brand.

There are many design competitions around the world, such as the BraunPreis, Core77, Design Intelligence Award (DIA), dmi:Design Value Awards, Fast Company Innovation by Design Award, Red Dot, and SEGD Global Design Awards. There are also many awards for specific areas of design such as the American Institute of Graphic Arts (AIGA) awards and the International Interior Design Association (IIDA) Design Awards.

In this section, I will discuss the criteria listed for five international design competitions from around the world as an indication of excellence and what can make a global product successful. The five international award competitions I will examine are:

- iF from Germany
- Good Design Awards from Australia
- IDEAs (International Design Excellence Awards) from the US
- Design for Asia Awards from Hong Kong
- INDEX Awards from Denmark

All of these competitive awards use a design jury made up of experts in the field. You will find that top juried international design competitions and their award criteria provide clues as to what products and services will be worthy of global greatness (Grand or Gold awards).

iF (Industrie Forum) from Germany

The iF Design Award has long become a symbol for excellent form, aesthetic quality, and user-focused, ergonomic, and efficient design in all disciplines by companies around the world.

Since 1953, the iF awards have followed six guiding principles:

- Identify, support, and promote good design
- Raise awareness of design among the public and the role it plays in our lives
- Help companies integrate design into their long-term strategies
- Safeguard the role of professional designer and boost awareness for this job profile
- Effect social change through design
- Support talented young people and create a public platform for young designers

While aesthetics and ethics ("good form") were initially the focus of the award, iF has articulated the requirement for user-focused, ergonomic, and efficient design. Products still need to display pleasing visual characteristics, but it is not the only criterion.[12]

Good Design Awards from Australia

Australia's Good Design Awards date to 1958. Good Design Australia, which manages the annual awards, is an international design promotion group, and it remains "committed to promoting the importance of design to business, industry, government and the general public and the critical role it plays in creating a better, safer and more prosperous world."[13]

Dr. Brandon Gien is the CEO of Good Design Australia and chair of Australia's Good Design Awards program. Gien hopes that design will remain human-centered even in the wake of technology that moves us into the future. He believes that designers are capable of problem solving at a much higher level than just designing products, and he thinks the design field will aspire in the future to embrace a wider role and help solve the world's biggest and most complex challenges.

Good Design believes that design is:

- A key driver of innovation, economic growth, export, and productivity
- An essential link between creativity, innovation, and commercial success
- The key to increasing social well-being and prosperity through empathy and insight[14]

For the Good Design Awards, all entries are evaluated based on three overarching criteria: good design, design innovation, and design impact. In addition, judges evaluate each entry based on specific criteria aligned to each of the 10 design disciplines represented in the awards.

The specific evaluation criteria for product design are outlined below:

Form

Is the design styled to be appealing and desirable for the intended market?

Is the design visually resolved and does it evoke an emotional connection?

Does the form convey the function and use unambiguously and intuitively?

Function

Does the product perform the function it was designed for?

Is the product easy to use and understand?

Is it ergonomically designed and does it advance the user experience?

Safety

Does the design protect the user from harm?

Does the design safeguard against unintentional use?

Does the design comply with all applicable standards and regulations?

Sustainability

Has the product been designed for disassembly and recyclability?

Is the product water, material, and/or energy efficient in its day-to-day use?

Do the materials and processes used have a minimal impact on the environment?

Quality

Has the product been well made and finished?

Does the quality of the product correspond with the desired price point?

Are the chosen materials and manufacturing techniques appropriate?

Commerciality

Does the product represent good value for money at the intended price point?

Is the design likely to increase the brand value of the company?

Is the design likely to result in a return on the investment made on design?

Innovation

Is the design new and original?

Does the design use new materials and technology in a clever way?

Does the design have any world-first features?[15]

IDEAs from IDSA

The International Design Excellence Awards (IDEAs) were started in 1980 by the Industrial Designer's Society of America. IDSA set up these global design awards to establish a benchmark for design and weather "shifting fads or transitory styling."[16] Gold winners for the IDEAs are housed in the permanent collection at the Henry Ford Museum in Dearborn, Michigan.

The criteria for this award are:

- Design innovation
- User experience
- Benefit to the client

- Benefit to society
- Appropriate aesthetics[17]

The user experience criterion focuses on "human-centered design," but there is also a need to benefit the client and society. Aesthetics, which used to be a key criterion, is now one of several equally important requirements.

Design for Asia Awards from Hong Kong

Since 2003, the Hong Kong Design Centre has held the Design for Asia Awards (DFA) that celebrate design excellence and acknowledge "outstanding designs with Asian perspectives."[18] Specific criteria for the awards are:

- Overall excellence
- Use of technology
- Impact in Asia
- Commercial success
- Societal success[19]

Criteria for DFA awards reflect a focus on culture (especially Asian culture), sustainability, and technology. These criteria reflect the desire to support and preserve Asian culture but to also remind those producing products in Asia to be aware of environmental and sustainability issues. The criterion of commercial success serves as a reminder to have a financially viable product for the market. The technology criterion encourages the use of the latest technology in product and service design. There are separate Grand Awards for issues such as culture, sustainability, and technology.[20]

INDEX Awards from Denmark

In 2002, Kigge Hvid and a team from Denmark started thinking about criteria for an award that was inclusive of products and services, while focusing on using design to improve life. This INDEX Award, first awarded in 2005, was one of the first to include service design.[21] The site for the INDEX Awards reads:

"And now, all design events, design conferences and design agencies around the world recognize the power of Design to Improve Life and allocate at least part of their program, work, or resources to advance this important agenda. The private sector has also followed suit, realizing the sizable commercial potential in designs that have the capacity to improve people's lives."[22]

Underpinning the INDEX Awards is the World Economic Forum's 17 Sustainable Development Goals (SDGs), discussed in the next section, which were put together by countries uniting to "guide us all toward living fair, fulfilling, and healthy lives."[23]

Global Sustainable Development Goals (SDGs)

In 2015, the United Nations (UN) member countries came together to create sustainable goals for products and services to reach by the year 2030.[24] These goals have raised the global profile of important topics like ethics, fairness, and environmental protection and have been supported by the World Economic Forum. The list below shows the goals that countries are encouraged to meet when introducing new products and services:

1. No poverty
2. Zero hunger
3. Good health and well-being
4. Quality education
5. Gender equality
6. Clean water and sanitation
7. Affordable and clean energy
8. Decent work and economic growth
9. Industry, innovation, and infrastructure
10. Reduced inequalities
11. Sustainable cities and communities
12. Responsible consumption and production
13. Climate action
14. Life below water
15. Life on land

16. Peace and justice, strong institutions
17. Partnerships for the goals

This set of goals is an ambitious undertaking, and the design community is taking a realistic look at what they can do to help achieve these goals. Incorporating these goals into design team discussions all over the world can increase the possibility of satisfying some (or hopefully all) of these goals.[25]

What Makes a Product Great?

In summary, when design teams consider new innovations for the global market, there are many issues to bring into the design process in order to produce viable solutions. Many of these issues will become more important in coming years as consumers demand more from their products and brands, thus affecting brand value and revenue. Listed below are some of the key goals for a great global product or service; these goals are taken from a cross-section of international design awards:

- Benefits to company (sells well, supports brand)
- Contributes to local economies (safe, fair employment, ecologically sustainable)
- Contributes to society or culture (is useful, helpful, and appropriate)
- Contributes by offering unique innovation and aesthetics (has appropriate styling, ease of use, and has unique or original aspects)

Companies and public organizations can evaluate their products and services against these goals. These criteria can serve as guidelines for making a product or service that contributes to the world in a meaningful way. For the purposes of this book, I will use the above listed criteria as the underpinnings of great global products.

What Does This Mean for the Future of Design?

The design process has expanded because consumer desire for design has expanded. It may seem overwhelming to bring a successful global product to market after reviewing the requirements listed earlier in this chapter, but the reality is that many organizations continue to move in the direction of improving the environment and society while simultaneously increasing their earnings and brand value. These best practices and goals will start to make their way into more product and service design briefs.

Cultural and societal missions of companies will continue into the future and even expand. In the future, companies may want employees to have an individual mission statement that contributes to the world. Service awards for employees have been around for decades but service work may become a more integral part of an employees' overall responsibilities. Only a handful of companies today allow their workers to go on service retreats. This may grow in the future as the workers request it as a part of their work package. Vacation/service retreat combinations may also be a way to increase a company's contribution to the world.

As a designer, you will need to explore the world in order to understand the context of complex problems ahead of you and your company. A designer usually seeks many unique experiences and places to visit in order to heighten their experiences and senses. This helps you to become better at design thinking and design reasoning, as these other stimuli may help you come up with innovative and novel solutions, especially when compared with just interacting with your team, or searching for information on your computer.

You will have a much larger scope of issues to address in the initial problem-solving or opportunity-generating project phases. This information can come in many forms—such as data collected, images from a culture, new legal requirements for products and services—all of it will need to be synthesized in order to glean insights for solutions.

The design teams may become large and diverse, and design managers and project leaders will need to organize them for best effects. Some team members might be needed in the earlier stages, when expertise is needed on materials use, cultural norms, or informal early research methods. Working with larger teams will bring more expert information to help with decision-making, but it can also alienate team members who do not feel they are part of a cohesive group.

The product and service goals listed in this chapter are used to satisfy customers, keep them positive about the brand, and protect communities, environments, and workers. In light of this, you will need to take into consideration the goals and aspirations of the companies and consumers when designing and developing a product. These product and service goals, when viewed together, may seem like a daunting task, but these goals can help fuel innovative thinking. The criteria presented in this chapter for good global products give the team a framework from which to work and then to compare their solutions. Designers will always try to satisfy as many of the criteria as possible for new design and innovation or product revision.

In the next chapter you will find 10 global impact factors I have established that may help (or hinder) designers from reaching their goals of making successful products, services, and experiences. At the very least, these factors will help designers and their teams to gather as much information as possible as a way to avoid risk. These factors relate to cultures, resources, governments, and more, and they need to be considered before launching products and services in other countries.

CHAPTER 2

Impacts on Product Success or Failure

When the *Economist* declared "data is the new oil," I do not believe they were specifically thinking about the future of design and global products.[1] However, having the right data is imperative; it can make decision-making much easier and significantly lower risks throughout the entire designing, manufacturing (or coding), marketing, and distribution phases. In addition, if companies are expected to do good things within a country or region, they may elicit a richer variety of opportunities from data gathering and analysis.

Artificial intelligence options and big data may become a major part of the front end of the design process in the future. By data, I mean information coming from various research projects, or data mining from other sources that might provide useful information. Proper data gathering and analysis can point to problems or opportunities and be far reaching by supporting or rejecting certain design ideas and insights. I consider a lack of adequate data before starting a project a weakness in the problem-solving process. Until vetted, or verified big data are widely available, we will have to depend on the current method of using computers to search for information pertinent to our interests through software alerts and search engines.

In Peter Frankopan's *The Silk Roads*, he discusses why cities became information hubs for early civilizations:

"There was good reason why the cultures, cities, and peoples who lived along the Silk Roads developed and advanced: as they traded and exchanged ideas, they learnt and borrowed from each other, stimulating further advances in philosophy, the sciences, language, and religion."[2]

Information traveled faster in condensed and connected parts of the world such as those on a Silk Road. Today there is a vast amount of data on the internet that can be easily accessed from many parts of the globe, but it still benefits companies to establish themselves where they can have access to the latest information from the latest influx of people. This is the modern Silk Road.

According to Eric Weiner in *The Geography of Genius*, certain areas of the globe have served as hotbeds of inspiration. For example, innovations in philosophy and government in Athens, Greece; finance and art in Florence, Italy; arts and science in Kolkata, India; industry and enlightenment in Edinburgh, Scotland; and music and psychology in Vienna, Austria.[3] New ideas grew rapidly in all of these places. Then as now, the mingling and sometimes clashing of ideas and people spurred new things. Artists, writers, poets, bankers, inventors, and others could watch and learn from each other. They inspired each other by creating art and music, and improving ideas and systems. Although all of the cities Weiner mentions are not on the Silk Roads, they were clusters of civilization where new ideas were born. Silicon Valley is one of the last places mentioned by Weiner, as a concentration of computing technology and specialized genius within a geographical area.[4]

Areas of Impact

In this chapter, I will provide a list of potential impacts on new products or services in relation to resources, social issues, and physical conditions—issues that affect product and service design. In order to be sure your team has done sufficient prior research before designing a product, a more intensive information gathering is necessary.

If you are marketing a product (especially if you are considering marketing in another country), now is the time to set up and track news feeds to detect trends, problems, and opportunities. An example of this would be if you were thinking about designing portable water filtration units for rural India, you might want to set up news alerts around this issue. These news alerts could be: filtered drinking water, India assistance societies, India politics in rural areas, or portable water testing units. This, coupled with a study to gain insights into the culture, history, geography, political climate, distribution and more, would provide a more thorough examination of the risks and opportunities involved.

For large, diverse countries and regions, you may need to set up several news feeds and historical studies. Large countries like China and India have disparate regions, so it would be a good idea to set up searches that cover their diversity. Large global companies often engage in fact-finding on the front end of their process, but their information-gathering methods tend to be narrowly driven when compared to the wider array of information that can and should be gathered. The company may ask questions such as "What farming equipment is being sold in Brazil today?" or "What are the legal implications of selling our products in Ecuador?" Of course, the companies do their research, but not all of this reaches the design team, where it can make a big difference in the design of the product. Companies will need much more complex information that is current, verified, and possibly pointing to trends in governance and cultures in the region where they are interested in selling products and services. They will need much more depth and breadth to their information and will need to share all of that with the design team.

All companies try to smooth the way for their brand to be sold in another country, but even major companies with employees working in different parts of the world have been surprised or thwarted by events that are out of their control. Natural disasters are common in certain geographical areas (typhoons in Asia, hurricanes in the Atlantic, earthquakes along the coasts of the Pacific Ocean). Political unrest is ongoing in many regions, and it can eliminate or hinder marketing efforts in a very short time. Cultural shifts in issues related to a product can occur quickly due to social media. An informed company can plan for some of these risks.

Emerging cultural shifts and changes in government policies may be the most difficult—but important—information to gather over time. Data gathered from different sources will be needed on a continuous basis over a long period to really understand a changing situation.

While not exhaustive, the impact factors presented here will provide companies with a guide for collecting information related to proposed

TABLE 2.1   10 Key Impact Factors for Global Products

Available Resources	Social Issues	Physical Conditions
1. Educated markets	5. Cultural acceptance	9. Climate and geography
2. Financial resources	6. Personal ownership	10. War and peace
3. Technology infrastructure	7. Government support	
4. Fertile ground for innovation	8. Individual and/or team passion	

product sales in new markets. These 10 factors are grouped into three areas (Table 2.1).

Available Resources

Available resources in the region where you would like to sell a product or service can be critical to success. It's much easier to move a product into a country that already has educated markets, financial resources, technology infrastructure, and a fertile ground for innovation. Plus, information may uncover a shortcoming in one or more of these areas, providing insights for new opportunities. Listed below are a few of the questions the knowledge-gathering team should ask.

1. Educated Markets

- What is the age of my target audience and how were they educated?
- Can they use computer technology?
- Do they understand the idea behind brands and brand loyalty?
- Is it a country that is just developing or changing its educational system?
- What would it take to teach them about my product or service?
- Is translation required?

If the region has a strength in a specialized area (such as France and food), it may be easier for you to engage in the market if your product or service relates to that strength. This would seemingly be easier than starting something in a region that has no interest in your product because such a region likely will be more educated about your product or service. The company selling the product or service will need more data to gauge the strength of the product acceptance or resistance.

Consider a partnership between École Polytechnique in Paris, France, and Carnegie Mellon University (CMU) in Pittsburgh, Pennsylvania, USA. The two institutions are conducting research and development in the area of artificial intelligence (AI) robotics for cooking and food delivery. In this effort, Sony Corporation and CMU will cooperate on AI and robotics research at the École Polytechnique in Paris. Sony's goal, according to Hiroaki Kitano, who serves as project lead, is to make "AI and robotics more familiar and accessible to the general public."[5] The question will be whether they can produce a process or product that the French will accept.

2. *Financial Resources*

- Are the people of a particular country or region wealthy enough to purchase the product or service you are trying to sell?
- Do they have personal funds saved or would they need to borrow money, especially for larger or expensive items such as a car or luxury product?
- Is there credit available in the country?
- What about the general wealth of the country or region (rural vs. city)?
- Can they maintain their physical infrastructure for distribution? Do they have a physical infrastructure for distribution?
- Are governments wealthy enough to support tax breaks or other enticements for entering their country or region?

Alibaba and Tencent, China's dominant e-commerce and mobile payment companies, are working to assist medical care in China. With a rapidly aging population and a lack of hospitals and doctors, they are building artificial diagnostic tools to help medical personnel. As an example, Jack Ma, cofounder of Alibaba, has invested in artificial intelligence tools for screening lungs.[6]

Wealthier regions or countries may find it easier to invent new products and services because they have the resources and the educated population. Even though you may have a smaller but viable company you may try to partner with those who are wealthier by offering a specific skill to assist their emerging technologies.

3. Technology Infrastructure

- How advanced is the technology in the country where you are trying to sell your software product or service?
- Is the internet speed adequate in the region of your target market?
- Is the number of networked users high, and does the bandwidth and infrastructure support the product?
- If it is a medical product that needs consistent electrical energy, does that country have a strong energy supply in remote areas?
- Are there stores and venues available with the necessary technology in place?
- Can you easily have your product included in local stores?
- Is there a culture where you need to sell to middlemen first before you reach your market?
- Are other items needed to support your product or service?
- Are there taxes, tariffs, or local systems in place that could inhibit the sale of your product or service?

Tusheti, a rural Georgian province on the border with Russia, is extremely remote, cold, and rugged. A new project is underway to bring internet to the region in 2018, with the hope that it will bring tourism and business development.[7] If a company specializes in unique tours or extreme sports, this information on bringing internet to a remote region can lead to new opportunities for these companies. Collecting data on new events around the world can lead to insights for new products and services.

Five months after Hurricane Maria wreaked havoc on the island of Puerto Rico in 2017, hundreds of thousands of Puerto Ricans were without power (and many still are as of this writing). Power companies that had gone there to fix the infrastructure have been pulling out as their contracts and budgets ended. Many of the areas left without power are in the mountainous regions where the working conditions to restore power were too difficult.[8] Understanding the plight of the Puerto Rican people and what their specific needs may be is considered seed for development of new products and services.

While disasters and suffering are unfortunate, there may be an opportunity to help the situation with your product or service. Your product might have been designed for just such an occasion or purpose. In addition, working with the latest disaster relief groups can help you to improve your products or expand the scope of what your company offers.

4. Fertile Ground for Innovation

- Does the country or region have the type of culture that is proactive, innovative, or creative?
- Is there an available variety of experts to help set up your business, and can they be hired if needed?
- Are there artists, designers, entrepreneurs, futurists, inventors, and other creative people in the culture?

When environmentalists in Australia concluded that shark nets were harmful to other wildlife in the sea, high-tech entrepreneurs there collaborated to design and develop the Westpac Little Ripper, a drone that uses hundreds of thousands of images to identify sharks. The accuracy is 90%, whereas accuracy by the naked eye is 20% to 30%.[9] This shark-identifying drone is one recent example of products that have emerged from a diverse group of experts working together.

China's science parks, art hubs, and evolving educational system are examples of a country setting up fertile ground for innovation. These changes are occurring mostly in the major cities of China, but some of these changes are affecting the rural populations as well.[10]

Social Issues

Awareness of social issues in a country or region where you would like to market your product or service can be just as important as available resources and physical impediments. Issues such as inhospitable cultural differences, hostile governments, lack of personal ownership of ideas, and different attitudes or work ethics can help or hinder your marketing efforts.

5. Cultural Acceptance

- Is the culture ready for your product or service?
- Is it a conservative or liberal culture, or a mix?
- Are women's rights supported? How much purchasing power do women have?
- Is there a strong sense of freedom of speech, or is it a restricted culture?
- Is religion prevalent, and could it be an issue related to your product?
- Is idea theft prevalent?
- Who are the early users of your product or service, and how can you reach them?

Cultural acceptance can shift quickly. In Chile, American companies such as Coca-Cola, Kellogg's, McDonald's, and PepsiCo are struggling to sell some of their products because of high sugar and fat content. A *New York Times* article, "Waging a Sweeping War on Obesity, Chile Slays Tony the Tiger," brings to light how changes can come rapidly to one society and then quickly spread to others. Brazil, Ecuador, and other countries in Latin America are looking at what's happening in Chile and may set up restrictions in relation to high sugar and fat content.[11]

For the companies mentioned in the article on Chile and the rise in obesity, it may seem like this issue suddenly emerged. But this topic likely had been already discussed in some of the local papers. Although the companies affected are large and have expertise in information gathering, they benefit from learning about this issue earlier rather than later. That would give them more time to respond and possibly alter their products or their sales approach and avoid damaging public relations.

In the United States, foreign automakers such as BMW, Nissan, and Toyota chose to build automotive plants in the southern regions, which are arguably less supportive of unionized labor than their northern counterparts. The Chinese automotive glass company Fuyao opened a plant in Moraine, Ohio, and is now concerned about possible unionization and what it could mean for the company. Bill Russo, a former chief executive of Chrysler China, said: "This is part of the growing pains of going global.

. . . You are going to have to deal with different cultures and different labor-management relations."[12]

A newly emerging cultural issue centers on addiction to technology. These concerns may spread globally, which could affect software products. In "Digital Addiction Stirs Worry Even in Its Creators," author Farhad Manjoo states, "Several former Facebook executives, the very people who set up the 'like'-based systems of digital addiction and manipulation that now rule much of online life, have begun to speak out in alarm about our slavishness to digital devices."[13] Of particular concern is the creation of the "short-term, dopamine-driven feedback loops" that are used in these programs.[14]

The concerns about digital addictions are not new. They have been surfacing in different countries and may one day be the impetus for legislation that restricts or bars addictive programs.

6. Personal Ownership

- Will the government allow you to own something if you create it, such as with a patent system?
- Will the government allow someone else to use it or support your claim of ownership?
- Are products and ideas often shared in this culture or region?

At various times in history people did not have personal ownership of property, businesses, or patents. They were owned by the state (or the king or emperor). Personal ownership can be an impetus for inventing new products and services and creating personal wealth, and those countries that allow personal ownership may have more eager entrepreneurs.

7. Government Support

- Does the government encourage new imports?
- Does the government require that you disclose your technology or trade secrets?
- Is there an effective patent system? Is the patent system respected?
- Is yours a product that the government might welcome, thwart, or copy?

- Do the national and local governments work well together?

Companies can look at governments in countries and regions and see opportunity or insurmountable problems, depending on what their product or service offers. In politically unstable areas, it is even more important to study the history of the region and current events.

As an example, lack of investment by a government in telecommunications in a country or region is problematic. It may be due to lack of funds, or possibly by design; some believe this is the case in Egypt, due to the Arab Spring and state-run monopolies. "In 2016, Egypt ranked 146 out of 150 countries for fixed broadband download speeds, according to Speedtest. The only worse country in North Africa was war-torn Libya."[15]

As a counter to those regions experiencing trouble, consider friendly Estonia, which is working to attract talent by providing "e-residency."[16] In his book *How to Fix the Future: Staying Human in the Digital Age*, Andrew Keen writes that "Estonia is the first country in the world to offer 'e-residency'—an electronic passport that offers any small businessperson the right to use legitimate Estonian legal or accounting online services and digital technologies."[17] Fingerprints, biometrics, and a private key are used to establish online identity. The leadership of Estonia felt that it had a disadvantage with its cold climate and geography, so the leaders decided to open up Estonia's borders in a virtual way, disrupting the idea of physical borders between territory and citizenship.

8. Individual and/or Team Passion

- Can you find the people with passion in other countries and regions to help you with your product or service?
- Can you find the right people for teams to support the development, manufacture, marketing, and distribution?
- Is loyalty valued in their culture?
- Will they protect your ideas while they are being explored?
- Do any of these people have a network they can contact to help you with your product development?
- Can they work effectively in teams or as individual contributors?
- Does there seem to be a good work ethic?

Jeff Bezos (Amazon), Richard Branson (Virgin), Bill Gates (Microsoft), Arianna Huffington (The Huffington Post), Steve Jobs (Apple), Jack Ma (Alibaba), Elon Musk (Tesla), and Ginni Rometty (IBM) inspired others with their passion, ultimately leading their respective companies to elite status. Their examples have inspired other entrepreneurs around the world to start their own companies or organizations. Individual passion, combined with the right team and the right work ethic, forms strong pillars of success.

Physical Conditions

Most natural disasters are unpredictable, but some technologies enable us to track and gauge the strength of hurricanes, typhoons or know the severity of an earthquake. This is not yet the case with floods or fires. However, scientists are working to learn when and where an earthquake might occur. Topological maps are helping scientists to understand the location and potential force of floods or the direction of a fire.

9. Climate and Geography

- Does the product you want to sell globally have adaptability in different climates or geographies?
- Is distribution a problem due to geography or climate?
- Are there any benefits or opportunities to extreme climate or geography that could help sell your product, such as extreme sports gear or safety equipment?

In southwest China in 2007, I visited with elders in the villages outside of Kunming. They told the story of Japanese businessmen who came to their region and encouraged some of the farmers to start growing a particular type of Japanese mushroom. The mushrooms were very expensive in Japan, and the reasoning was that they could grow them cheaper in China and make a better profit. Some of the Chinese farmers took out loans to buy what was needed to grow the mushrooms. When the mushrooms died, the farmers and the Japanese businessmen realized the mushrooms could not adapt to the climate and geography in southwest China, as they had in Japan. This left many farmers in debt from the failed project. Worse, it eroded the trust that the farmers had in new opportunities. This example illustrates why social responsibility and familiarity with a region are critical in economic situations—and what can go wrong when familiarity is lacking.

10. War and Peace

- Is the country or region in a state of perpetual war or unrest?
- Does the government turn over rapidly?
- Are there other countries threatening that particular country or region?
- Are major changes anticipated in their alliances?

In 2017 the French company Lafarge, which has a cement plant in Syria, found itself the subject of a criminal investigation and a civil lawsuit in relation to its evacuation methods in a crisis. As fighting factions neared the cement plant, workers heard explosions. They fled in two small cars and a delivery van.[18] Companies cannot always know when their employees are in trouble, but real-time information gathering helps companies to assess risk and possibly make timely and critical decisions. In the case of Lafarge, real-time alerts on the fighting in Syria would have helped them better understand the risk.

In 2018, the Kenyan government disconnected four private television stations that were broadcasting a protest. The Kenya Human Rights Commission believed the "government's action was moving down a worrying path."[19] A disruption in advertising during the shutdown could be problematic for a business in that country. And problems could multiply in the event of larger protests.

Again, gathering information from the region over a period of time will help a company decide whether to invest their time and energy there. Information is key to risk mitigation, whether there is an opportunity or a problem.

How to Use These Factors

Companies or individuals that want to market globally can use these 10 impact factors to guide their information gathering. Today, information gathering can be as simple as using Google alerts, or asking a coworker to track information in a particular region. This information, combined with vetted, verified data, will help determine whether you want to commence with projects for a particular region or country.

There might be great promise in terms of educated markets, financial resources, and cultural acceptance in an area where you would like to launch a product or service. However, one major negative impact factor, such as a government thwarting particular imports, can dampen any excitement about new markets.

It may also take even more varied types of information gathering to learn about topics that are not in the news, such as looming social or political issues. Content from local blogs or newspapers might reveal that there is a problem (or opportunity) with the lack of an educated market or product distribution system. Some of the most important information may be in the country's national language and not accessible to English-only audiences. This may require translation services, but these translations can provide huge insights into a culture. A note of caution, however: zealous information-gathering requests should not be misconstrued as condoning spying or as an excuse for stealing corporate secrets. It must be made clear to any employee who is searching for information to aid their company that accessing a data base or system that is not open for public consumption is inappropriate.

Ideally, information gathering should occur before the design process begins. Any concerns that arise can become a part of the design problem, or you may even uncover an opportunity. For example, if a company is selling a product for children and identifies inadequate education in a region, the company may try to redesign their product to include an educational component. Childrens' clothing or blankets might have numbers or alphabet letters on the fabric to help a very young child become familiar with those shapes.

Many large global companies have teams in place to do the research to make a sound decision, but smaller companies might not have those same resources. Individuals or smaller companies can investigate most of these 10 impact factors on their own, which might lessen the expense of legal or other expertise.

This is not an exhaustive list of impact factors, but they provide the broad areas that need to be addressed in order to globally market your product.

What Does This Mean for the Future of Design and Global Products and Services?

If data is properly gathered (meaning it is properly vetted and organized), it can be an incredible asset to companies and designers. For designers, this means that team size and makeup can vary greatly from the beginning of the product development process to the end. More diversified team members may be needed up front in the design process, such as sociologists, economists, marketers, international lawyers for tax and trade, and cultural representatives from the region or area of interest. This early influx of information and data to the design process can be very inspiring for the team, as there will be more information points to consider. This may increase time spent during the information gathering phase, but it can also result in a richer array of potential solutions.

Companies and designers may find themselves working with independent data validation companies, depending on the product. Jean-Luc de Buman of the Geneva-based SGS Company says: "Even if technology appears to remove the need for independent verification, there is still a role for giving external assistance to the public."[20] Buman believes so much data is provided by suppliers directly to companies that someone should independently verify data. Claims about safety, materials, chemical use, or any other questionable issues should be verified. Verification of data, news, and other information coming from another source may be needed today, especially in light of online information that may not be true.

As such, designers will need to become more familiar with statistics. Understanding statistics, and how they can be manipulated, is an important skill. A fact may seem important, but then you find the sample size was very small, or the right questions were not asked. If you are going to work with more data, you may need to help companies frame the information you need. When design research methods became popular in the 1980s, most designers did not know how to apply the information that anthropologists and sociologists would collect for them. But once designers began working with the anthropologists, psychologists, and sociologists to help frame the research, they began to get usable data.

CHAPTER 3

The Design Process, Design Thinking, and Innovation

Creative Problem Solving

There are many paths to creative problem solving, which takes place in every area of study. The arts, business, humanities, and sciences have their own methods, processes, and terminologies that have evolved with each those disciplines. Around the world, today's artists, scientists, business-people, and other professionals engage in creative problem solving, but not necessarily within a design process.

A computer programmer may try to solve issues with a difficult piece of code but may not be a part of a larger design process. A biomedical engineer working in her lab may try to find a cure for a disease, but she may not be involved in a wider design process. In some cases, an inno-vator's own interests and passions lead them through their own creative problem-solving process, which may be different from the design process.

Even though the design fields are fairly young in comparison to other creative disciplines such as art and architecture, I include design under the umbrella of creative problem solving. The field of design was first inspired by arts and crafts methods. Initially, design was artistic and intuitive and did not involve solving major problems, other than to make a desirable and functional object. Now, design, inspired by science and technology, has taken on much more complex problems to solve. Even a simple problem in the past, such as making a functional tool, has grown to be more than just the tool. It represents a brand, it may have technology embedded, it must be safe, it must meet legal requirements, and be made of durable but sustainable materials. The simple act of making coffee with an electric coffee pot has grown more complex over the years. No more plug-in percolators; we now have coffee "systems" in our homes that can make an expresso, latte, or "Americano" style of coffee and froth our milk with the same system.

More than 50 years ago, Christopher Alexander, a British designer, realized that design problems were increasing in complexity and number.[1] He listed four reasons why the design process needed its own methodology, and those reasons are just as relevant today:

- Design problems had become too complex to treat them purely intuitively.
- The amount of data required to solve design problems had increased so rapidly that one designer could not collect—let alone process—them all.
- The number of design problems had increased rapidly.
- Totally new design problems were emerging at a faster rate than previously, so ever fewer design problems could be resolved by referring back to long-established practice.[2]

Let's explore the design process that will continue to be refined as we move into the future. You will still have the lone designer working on a problem they want to solve or something they want to invent, and many of these people can be called entrepreneurs. However, design problems may need much larger teams with more diverse members to solve the number of complex problems emerging today.

The Design Process

Design can be called systematic in that there is a process to cycle through in order to reach an effective design solution. The design process, along with design thinking (and reasoning, which will be discussed later in this chapter), can bring you closer to insights and innovation.

TABLE 3.1 The Design Process

Stage 1 →	Stage II →	Stage III
Problem or opportunity inquiry	Solution generation	Solution evaluation

Table 3.1 shows the simplified stages that the individual or team progresses through. Although the stages are represented in a linear way, the stages can vary in duration and effort. They may be repeated, if necessary.

The design process is distinguished from other creative problem-solving processes in the arts and sciences because the experience of the customer is often a starting point. Very often, art and design influences are mingled in a project, or science and design influences are mixed, and the disciplines can inform each other with fabulous results. The gift of the arts to the design field is self-expression and the gift of science to the design field is knowledge.

As an example, someone who handcrafts a chair may focus on their expression of a chair rather than the science of ergonomic seating. However, someone designing seating for an airport must study proper use, safety, ergonomics, materials, manufacturing, and even cultural habits.

The design process sets up a positive (yet critical) thinking experience that encourages innovation. It also has the built-in goal of making a product or service improved for humankind and the environment.

First Stage: Problem or Opportunity Inquiry

The initial phase of the design process gathers both general and specific information that will help the design team solve the problem. The following questions might be asked during this first stage of the process:

- Who is the competition?
- If there is a competitor, or many competitors, what are they producing and why?
- What are their successes?
- What are their customer reviews saying?

- What new materials or processes can be used to give a competitive edge or be more sustainable?
- Where and to whom will the product be sold?

Designers will sometimes work with consumer researchers, ethnographers, historians, psychologists, and sociologists to learn more about the countries, customers, and cultures in which the product or service will be sold. Traditional research methods, such as focus groups, interviews, observation, or surveys, are used in this information-gathering phase. Designers take the information gathered during the first part of the process and find inspiration from the people who will need the product or service. They talk to consumers to gain insights and check the problem statement accuracy before starting the second phase (ideation) of the process.

This first stage enables the designer to examine the scope of the project. In these early stages, the Ten Impact Factors from chapter 2 should be used as points of inquiry in order to reduce risk.

For example, a major lighting company that produced bulbs was suffering from in-store losses due to bulb breakage. The brief, or problem statement, provided to the design team asked them to design a more durable lightbulb. The team first decided to make observations to understand why so many bulbs were breaking. Placing unobtrusive cameras focused on an in-store lightbulb display, the design team noticed customers picking up the lightbulb package in order to read the wattage, which was printed in very small type. This act—picking up the package—led to people dropping the package and breaking the lightbulbs. After this discovery, the designers increased the size of the wattage information on the packaging so a customer could read it from several steps away. This led to less handling of the package and, in turn, less breakage of lightbulbs. While the original problem statement may have led to a very expensive redesign of the lightbulb, the design process of problem inquiry in the early stages led to a quick and cost-effective solution.

Second Stage: Solution Generation

The second stage of the design process, Solution Generation, is when the designer considers all of the data in an attempt to gain insights that will solve the problem or confirm an opportunity.

The designer synthesizes large amounts of visual, verbal, and written information to begin generating a solution. This stage often appears mysterious to those not involved in the design process, so designers will express their ideas with quick sketches, simple computer animations, rapidly made models, or working prototypes in order to communicate their ideas. Computer software now plays a large role in this stage of the design process.

In the future, I believe that AI programs will enable the designer to enter design parameters and receive design solutions. AI can uncover issues and also provide data that can justify some of the solutions. The designer could become more of a steward of the process and curator of final solutions.

Software companies such as Adobe, Autodesk, and Microsoft have made continuous improvements to the functionality of their software, allowing designers to visualize solutions. 3D printing options have improved through better software, hardware, and print materials, and they can provide shortened times for prototype completion. Software to mock up interfaces for testing has also improved, especially in the areas of AI, augmented reality, and virtual reality.

Design Thinking (and Reasoning)

Design thinking can occur throughout the design process. It may change in focus at various points, but overall it is meant to be a flexible, positive, and critical process. The question "Why?" is often used throughout the design thinking process in relation to a particular product or service attribute. Asking this question forces the designer and design team to verbalize reasoning and decision-making.

The early stage of the design process is often referred to as the "fuzzy front end" of product development because it is not always clear where the process will take you. This is not necessarily a bad thing. Designers have learned to tolerate, and even embrace, ambiguity while they are problem solving. Tucker Viemeister, a creative legend in the design field, believes the ability of designers to "hold contradicting thoughts" while thinking of new solutions is one of the mental skills that aid designers and has served the field well over the years.[3] This ambiguousness while problem solving is something that other disciplines may find frustrating or difficult to do, especially if their discipline requires linear thinking.

Instead, the ambiguity of the problem-solving process can help designers synthesize disparate pieces of information. However, just having an insight is not the solution—it has to be applied to satisfy the problem. Designers have the ability to imagine one solution, and then imagine opposing solutions. An example of this might be when an interior designer is imagining an interior with materials that provide a calming experience, or designing the same interior with materials that provide an exciting experience. Designers can imagine these scenarios and then put them down on paper or in a computer program (visualization) so others can see how those ideas might be put to use.

Part of the skill in design thinking is to be *aware* of your thoughts throughout the design process, and not have them overly tempered by old beliefs or methods. It is a positive, highly conscious type of thinking that will open up solutions at the beginning stages of the process.

Design Reasoning

In *Creativity: Flow and the Psychology of Discovery and Invention*, Mihaly Csikszentmihalyi discusses the dichotomies that can occur within creative personalities: they often use convergent and divergent thinking or focus "like a laser beam" when necessary, and then may seem idle when they are really recharging their batteries.[4] *Divergent thinking* is when you open up to many avenues and mixes of information for potential design solutions. *Convergent thinking* is when you are narrowing your options for potential solutions. *Design reasoning*, or the type of thinking that switches between divergent and convergent thinking, sometimes occurs more rapidly than can be captured, or even consciously recalled by the designer or the team.

Design reasoning can seem like an endless number of correlations among data. Combining and comparing thoughts during design reasoning ("this with that" or "this with those") increases the number of possible solutions. The designer's thoughts may circle back over information again and again, looking to make a new or unique connection that solves the problem. This reasoning during the ideation stage is often supported by visuals; many designers sketch while reasoning takes place.

You may need some courage during the design concept phase if you are asked to explain your reasoning around your proposed solution. Csikszentmihalyi describes how the creative individual has a "willingness to take risks, to break with the safety of tradition."[5] Designers repeatedly summon the courage to think differently and present unusual or possibly controversial ideas—even when the business culture or society does not welcome them. It is especially tenuous in political business situations, where breaking with tradition might adversely impact another person, team, or manufacturing plant. Designers suggest change, and change is not always welcome.

Innovation

Innovative ideas can emerge at any stage in the design process. An insight into a possibility or a connection between two or more things can spark innovation. Desperate need or other life situations can be an emotional driving force behind innovations.

Innovations can be major or incremental. For example, although the wicks and wax of candles were improved throughout the centuries, electric light represented an entire leap of technology. The same holds true for the horse and buggy, which was continually improved but did not directly lead to the invention of the automobile.

One example of global innovation, the lowly paper clip, went through many iterations on both sides of the Atlantic in the early 1900s. People in different parts of the world generated ideas for holding papers together. Among them were a simple pin that pierced the pages (and the user's fingers), a T-shaped pin (which also pierced the fingers), and eventually the rounded Gem paperclip. This major leap emerged when the company Gem invented a machine that manufactured the looping paperclip. The Gem paperclip did not hurt fingers or pierce the paper, and it became a global success.[6] Yet, paper clips are not immune from future innovation. They probably have already seen a reduction in demand, and they may go the way of paper in the future as more files are stored digitally.

In the design process, you usually expect innovations to occur during the second stage, Solution Generation. However, insights that spark innovation can occur at any stage in the design process. New information may be added, whether it is from data, comments from users, or the availability of new materials, research, and technologies.

Whether you are engaged in thinking or doing, active engagement with the problem is necessary for great breakthroughs or innovations. In *The Art of Innovation*, Tom Kelley tells us that "whether it's art, science, technology, or business, inspiration often comes from being close to the action."[7] A designer or inventor must be involved with the problem and process to be inspired. The design process today is an immersive, action-oriented experience, usually involving engineers, manufacturing experts, social scientists, marketers, designers, and whoever else is relevant to the problem. The designer must stay both positive and critical of potential solutions or innovations.

As an example, if you are designing a new check-in service at the hospital for senior patients with no relatives, the design team would talk with that audience. The team would empathize with the older patient to see what could be changed to reduce the patient's anxiety and make the experience a comfortable one. Maybe the innovation is as simple as a person in a new position serving as a greeter and advocate—someone who stays with the patient during the entire check-in process. Or maybe registration begins at home with a new piece of software that allows others to input information. Or, after the design team talks with the patients, the designers may become inspired to create a new system for all registrants using the latest advances in imaging identification and personal virtual reality experiences.

When automated teller machines (ATMs) were invented, most people were accustomed to interacting with human bank tellers during limited banking hours. For people who needed to retrieve cash from their accounts at any hour, the ATM was a wonderful innovation. However, some customers missed the human interaction and were afraid or skeptical of the technology, wondering if they could make a mistake that would transfer their money elsewhere. Anxiety with ATMs again increased slightly when checks could be deposited directly into accounts through an ATM. My first try at that experience caused me to wonder: would my check *really* be deposited in the right account? Eventually, along with those continual improvements in functions and service, came trust. Many of these ATM service designs were made with the use of design teams that conducted research with bank customers and ATM users.

However, time and innovation march on. ATMs may be gone soon as we move to more online banking and purchasing with our phones. People have already developed trust (and practice) with electronic transactions, so the move to a smart phone transaction is not a major change. What may be a future disruption is the phasing out of cash. China today is almost a cashless society. They have also leapfrogged over the West's obsession with credit cards and instead transfer money to their cards to use as cash. Today's sharing societies have accelerated the need for easy and fast financial transactions. Today in China you can reserve a parking spot, pay your restaurant bill, loan money to a friend, and more with your phone and a few apps. These types of transactions will proliferate in other cultures as well, making it necessary to design transactions for safety and trust.

If society becomes cashless, you may look back and think it was an innovation leap, when in fact it was a series of minor improvements and innovations. Will you still need a wallet? Probably not. Wallets are used to carry cash, credit cards, and identification. They may go the way of the watch and be used as a fashion statement.

Another example of major global innovation is autonomous driving. It is likely to be achieved through a series of incremental innovations from around the world, which ultimately will result in a driverless car. Such cars already exist, but hurdles remain. Cultural acceptance, more advanced technology around safety, new laws, and successful business models, etc. are needed.

Not all innovation is technologically based. It is important for the designer to look at nontechnical solutions as well as technical solutions. Agricultural methods for farming, new ways to create art, or hand washing to prevent disease are examples of nontechnical innovations. The more recent idea of microloans for low-income people is considered a social innovation. In a LinkedIn post, blog writer Luis Rajas Fernandez cites innovative business models such as Gillette introducing disposable razor blades.[8] Refillable products such as cameras and film, vacuums and bags, printers and ink, and even candy dispensers became business innovations.

Third Stage: Solution Evaluation

The third and last stage of the design process is Solution Evaluation. Examples of the types of questions asked during this phase are:

- Is the solution suitable for the original problem?
- Is this solution really solving a problem, or is it an unnecessary product or service?
- Does the design have the intended effect for the client or company?

- Does it satisfy most, if not all, of the requirements for the new product or service?
- Can it be improved?
- Does the original problem need to be readdressed based on what has been learned?

Design research is used at this last stage to glean customer or user feedback through observation, user testing, or surveys. The data gathered from customer and user feedback is often added into the design team's evaluation of the product. If they feel their solutions were not successful, the team may go back to find other solutions or opportunities.

During this stage, prototypes are made. The sophistication of prototypes or virtual experiences will allow for more exacting scientific user feedback through sensory technologies. You will not have to ask, "How do you feel about this product or service?" You may be able to tell from sensory data collected from individuals.

Ethics and user research will become major issues in the future for sensory technologies. For example, if someone is fitted with sensors and their data (such as pupil dilation, heartbeat, or blood pressure) is collected when they are shown a particular image, it would be highly unethical to release that data if the subject is unaware their data is unknowingly accessed.

Unfortunately, the third stage of the design process is sometimes shortened by a company's desire to rush the product or service to market. After the product or service enters the market, sales numbers are often used as the main metric of success or failure. But a more robust evaluation is needed, including quality of manufacturing, ease of use, and success of marketing campaigns, the success of the design process, or any other pertinent information. Such issues can greatly affect the success of the next product. It is also good for the design team to reflect on what processes worked well and what might need to be changed or improved for future projects.

Design Research Informed and Backed by Big Data

It is entirely possible in the future that insights from big data will help to inform the problem statements given to the design team. Research helps designers better understand the design problem, and having more data and information will help designers to generate more viable solutions. Some common research methods that help designers understand the consumer or user include:

- Observation
- Interviewing
- Focus groups
- Participatory research

In the third stage of the design process, Solution Evaluation, design research helps with the assessment of ideas. These evaluations often uncover problems in the original design, which may necessitate adjusting the design brief. Many of the traditional research methods used in the West (such as interviews or focus groups) may need to be adapted to other cultures. While Western consumers are accustomed to surveys and focus groups, such methods are not as readily accepted in some cultures.

An example of this occurred during a software interface evaluation in Hong Kong in 2005, where a major US software company wanted information from teens about its new interface design. Focus groups of male and female teenagers were set up. The design team, of which I was a part, quickly realized the teens were too shy to speak in front of each other, so we separated the groups by gender. But even with gender separation, the teens were too shy to offer much insight into whether the interface was acceptable. What was suggested in the end was a modified "focus group/interview" method I called "bring a friend." We found that teens who brought a trusted friend with them were more open and instructive about changes that could be made to the interface. Some friends even disagreed with each other and provided more data during the session than was hoped for initially.

For more than 150 years Western consumers have become familiar with marketing and evaluation methods, but the Western culture of individuality and risk taking is not universal. Designers will need to become more sensitive to the nuances of culture and research, and possibly modify their methods while still upholding good research practices.

What Does This Mean for the Future of Design?

The role of design in the future will expand because the complexity of design problems has increased. Design as a process in the future will change as more experts are invited to weigh in, more consumers are involved, and new methods of capturing research and insights become available.

The design process, design thinking, and innovation will all be affected by big data and artificial intelligence. Although new technologies will be helpful and may increase available information, they may also complicate decision-making.

At the global level, product innovation (and therefore design) will be affected by the speed at which ideas circulate, and the ability of cultures to engage in innovation. Some countries may not have the resources, talent, or infrastructure to work on major innovations, choosing instead to focus on other priorities, such as feeding communities or fighting disease.

Big data and modified research methods will give designers, design teams, and innovators more information to apply toward solutions. Artificial intelligence will help to synthesize data and recognize patterns, especially in the medical and legal fields. Designers will need to assess the big data assumptions and work with data verifiers. Hacking and fake data will, unfortunately, continue to pose a serious risk, although we will likely see new software that will detect and curtail these types of activities.

Designers and companies may begin to include smart software involved in solution generation. Existing software can render an endless supply of shapes, textures, and colors for aesthetic evaluation for animators, communication designers, digital media designers, and product designers. For services, AI may generate text and visual solutions for mockups or models for the team to assess. These changes may not be welcome by existing designers, but the newly trained designers will automatically become familiar with these new technologies.

Capturing the nuances of design reasoning for other stakeholders will be important. Designers can capture and explain some of the insights that lead to solutions, but it is difficult to capture all of the points or the forks in thinking because design reasoning can move very quickly. It is best to document the process in any way possible, which enables designers to return to certain points in the process, recall salient points, or determine where the reasoning took another direction.

Designers will need to conduct user research in a variety of cultures that may not understand or respond to Western research methods. Alternative research methods may be needed, but they will still require ethics, rigor, and the other hallmarks of good research practice. Working with professional researchers is strongly recommended so your methods adhere to ethical guidelines and best practices.

Global product innovation could, in some cases, become more of a global "team" particularly for large-scope, multifaceted challenges, such as pursuing commercial space travel, saving environments, or feeding developing countries. You may see more design teams working on the same problem in different parts of the world for one company or organization.

Regardless of the challenge, society will still need beautiful, sustainable solutions for our problems large and small.

CHAPTER 4

Attributes of Great Product Design

If you would have asked a designer in the past what the most important aspect of a product is, you would have heard "it's the aesthetics; it has to look good or it won't sell." The physical attributes of color, shape, texture, or composition would have been most important. Another designer might say, "It has to function really well and be easy to use, employ ergonomics," or it is about "sustainability, price, and expert manufacturing."

Today's designers will say it is all of the above and more. Some of these characteristics will be more important than others, depending on what is being designed, but what will not change is the designers' goal of getting the details right for consumers and users.

Design Attributes

Candy Crush, iPads, Teslas, bubble tea, Star Wars movies, Nike shoes, and Xiaoice (a surrogate friend[1])—what makes these disparate products and services so popular in large markets? What do they have in common? People have experiences with products and emotional responses to them. The competitive game, the sleek input device, the eco-friendly car, the funny drink, the hero action film, the attractive sports shoe, and the AI friend all provide positive experiences and emotions, which make for satisfied consumers. Of course, this is a generalization, but experiences

TABLE 4.1   Product Attributes from Cagan and Vogel

Emotion	The feelings and emotions that emerge in relation to the product or service
Aesthetics	The sensory perceptions from visual, verbal, tactile, auditory, olfactory, and gustatory experiences in relation to a product or service
Identity	The product or service expression of personality, context, and timing—how it relates to the consumer
Ergonomics	The ease of use, safety, and comfort of the product or service
Impact	The company's ethical, social, and environmental actions, and how it relates to the consumer
Core technology	The reliability and enabling aspects of the technology used for the product
Quality	The craftsmanship, manufacturing, and durability of the product or service[2]

with products and services bring you into the present moment of life and heighten awareness through the senses.

To drill down further into the experiences, I will discuss the attributes to which people respond. In their book *Creating Breakthrough Products*, authors Jonathan Cagan and Craig Vogel explain that product value is broken down into attributes that make up a product's usefulness, usability, and desirability. They list categories of attributes (see Table 4.1) that they consider value opportunities: emotion, aesthetics, identity, ergonomics, impact, core technology, and quality—each contributes to the overall experience of the product.[3]

Cultural Beliefs and Experiences

A value assessment of products and services has a lot to do with innate beliefs, many of which come from the surrounding culture, government, family, or friends. Whether someone likes or dislikes a product or service can reflect their beliefs. When you experience a product or service, you give it meaning and an emotional connection.

Feelings about products are great clues into values and beliefs. For some, owning luxury items might make them feel uncomfortable. For others, owning luxury items makes them feel good. When you see someone with a luxury item, are you happy for them, or do you feel judgmental? Do you wonder why they would want such an expensive watch? You may be uncomfortable wearing such an expensive watch, or believe it is wasteful. How about owning an expensive car? Or an expensive house? For some, luxury may entail having free time and not involve a purchase at all.

When Henry Ford allegedly said, "Give them any color they want as long as it is black," he was thinking of manufacturing efficiency, and that black paint dried the fastest. However, the automobile went on to become one of the greatest expressive and emotional products in the world, with new styles and colors emerging every year. When the sleek and sophisticated Jaguar brand from the United Kingdom was purchased in 1990 by the US-based Ford Motor Company, the design of the car changed. It became muscular and powerful. This focus on muscles and power seemed to reflect American culture at the time, and that styling could be found on other products such as portable sound systems, motorcycles, and sporting equipment.

Automobile advertising campaigns often give clues as to whether the owner might be "carefree" and own a Chrysler convertible, or more of a "family man" with a Ford van, or "ram tough" with a Dodge truck. Advertising for future autonomous vehicles will have to take a new turn to entice passengers. Such advertising is likely to reference safety, convenience, cost savings, or the luxury of free time while you are riding.

Great Product Design and Added Value

Great product design can range from low-tech, low-cost housewares to highly scientific lab equipment. It can range from an inexpensive piece of basic clothing to an expensive space suit. Whether the product is amusing or useful, it must be well designed and give the user an outstanding experience.

However, moving products from one culture to another culture without studying the differences can be risky. What is acceptable in one country may not be welcome in another, and you can risk alienating or offending potential customers. References to "lighter is better" in ads from a German beer company did not go well with some viewers. H&M brand in the UK recalled shirts that said, "The coolest monkey in the jungle" after ads depicting a small black child wearing the shirt were deemed offensive.[4] H&M was quoted as saying they believe the fast pace of the fashion world allows questionable products to go unnoticed. Going forward, the company plans to use new artificial intelligence programs to screen their products for potentially offensive slogans or words.

If you buy a pair of Tom's Shoes and know that a similar pair has been sent to help someone in an impoverished village, you may like Tom's Shoes a little better than other brands that do not contribute to good causes. However, the shoes also must look good, feel good, and be manufactured in a way that minimizes harm to the environment or workers. Since Tom's Shoes has a mission to help others, you trust Tom's Shoes as a brand and want to support them.[5] It makes you feel as though you are giving back to the world through your purchases.

Clients, Customers, and Users

When customers buy products for themselves, such as a piece of furniture, they may ask themselves questions such as:

- Do I like the look of this?
- Can I afford it?
- Will it take the wear and tear of my family and pets?
- Is it comfortable?
- Does it fit my lifestyle, etc.?
- Is the company a "good company?" (meaning do they give back to a community or pay adequate wages)

As stated earlier, part of the emotional appeal today might involve the ethics of the company producing the brand. Is the company giving back, caring about the environment, and treating their workers in foreign countries with dignity? These aspects will become even more important in the future as companies build their brands.

Parents buying things for their baby serve as an example of *customers and users*. They are buying items for the baby to use. However, that new baby pacifier is probably designed to appeal to both the parents and the baby, such as glow-in-the-dark materials that make it easy for the parents to find the pacifier in the middle of the night. The clip attachment on the pacifier helps keep the pacifier with the baby, who is the user. The baby likes the naturally-shaped nipple on the pacifier.

The terms "user" and "end user" became popular when software became widely available and user testing began. Software designers were interested in the "user experience" and employed observational and interviewing techniques to assess the software for ease of use and satisfaction. User satisfaction would hopefully lead to an overall positive user experience.

You can have a variety of experiences, and reactions, involving entertainment, medical care, and travel. You hope an entertainment experience will be pleasurable. And while you do not expect medical to be pleasurable, you would still want a tolerable experience. You dislike airport ticket kiosks and parking garage exits that do not work. The frustration levels increase if you are short on time or feel held back. These problems open up opportunities for companies and designers to innovate and improve these products and services.

Most products and services available today have endured the launch phase and then undergone refinement through a redesign phase. Feedback from consumer and user reviews can prompt a redesign or refinement. Successful hardware (appliances, autos, camping gear, computers, household items, furniture, medical devices, etc.), software (games, video editing, taxes, etc.), or services (banking, delivery, education, hospitals, restaurants, retail, travel, etc.) have all been redesigned many times.

In the future, data gathered from global purchasing patterns and consumer reviews will likely provide design teams clues into different cultures. Companies or organizations that have this data will have the advantage of knowing what people are buying, what they are saying about a product or service, and how much they are willing to pay. In China, Alibaba is using its data to determine what customers might want as a grocery store experience. Shopping and paying with your phone in the grocery store or having your food cooked and delivered are aspects of this customer-centric enterprise.[6] In the United States, Amazon is using data to develop its own products that will compete with other brands that the company sells online. Amazon does have an advantage in that the company can see the number of sales and prices of the other brands being sold.

Are You in the Target Market?

If you find yourself annoyed or not enjoying an experience with a product or service, you might be having a bad day, or *you might not be the target audience*.

The next time you consider buying an item, ask yourself if you are part of the target market, or group, for which the product was designed. If you are, chances are the product has been designed to appeal to your values and emotions in some way. Products we like can create positive feelings or reduce negative feelings. Products and services can help us feel smart, wealthy, stylish, attractive, practical, prudent, industrious, and helpful. A day at the spa rejuvenates you because the services are designed to pamper you and help you relax. A simple garden tool could be your favorite product, and even though it means work, it was given to you by your grandfather, imbuing it with more meaning and value.

Everyone has experienced products they were not fond of, whether it was a gift that was not quite right or something that was outgrown. Maybe you do not want to support the product or service because you do not align with the company beliefs. Negative emotions often cause people to reject products and services.

In the future, products will be more expertly targeted to particular markets by virtue of more data and testing available. The target markets for products and services will then generate data, which may indicate potential success or failure in the second generation of products. Some products may have a much shorter lifespan if the information on sales, which can now be collected in real time, shows low activity.

Classic Product Attributes

Clients and consumers often ask what makes a product timeless or "a classic." When one thinks of classic products, brands such as Cartier, Chanel, and Louis Vuitton may come to mind. These companies have built their brands over many decades, and they are known for good design and quality of manufacturing. Classic products stand the test of time; they are acceptable in many cultures, maintain value, and often become recognizable for certain features.

Classic designs usually have some distinctive features that provide visual continuity, such as a mark, logo, or distinct material. Cartier watches are known for the style of their watch face. Chanel products are simple and bold, with two interlocking Cs. Louis Vuitton has its L and V mark.

In 1985, Swatch watches introduced visually engaging new designs and materials, and they were considered a comparably inexpensive watch. Older Swatches are now collector's items, and some of the original designs are now considered classic. While Swatches are still inexpensive compared to many of the classic watch brands, such as Cartier and Rolex, they have built a brand with consistently good design.

The Coach company provides an interesting exception to the classic rule of building the same consistent look for a brand. Coach had achieved success with classic bags, which became a favorite of professionals. The company became known for its sturdy, black and brown leather, clean lines, strong stitching, and brass hardware. In 2010 Coach redesigned a line of bags to appeal to a younger, global audience, and this colorful line became highly successful. The new Coach bag, with new branding, appealed to a younger, trendier audience. Coach still carries some classic bags but has moved on to other designs.[7]

Even fairly new media products such as Japanese anime have become globally popular since the 1980s by providing emotional themes and universal storytelling. Some anime, such as Doraemon, Pokémon, Sailor Moon, and Dragon Ball Z, might be classified as "classic." Sports memorabilia and other collectibles can become classic if they are widely known and treasured over time. If you look through Cagan and Vogel's attributes individually in Table 4.1, you can see that certain attributes such as aesthetics or identity will underlie aspects of a classic product.

The Future of Products and Services

Great product design can change over time. Products that you admire today may seem simple or naive in the future. Products and services in the future may connect to the internet of things (IoT)—objects that will be able to interact with each other or be linked to the web. Future products will have more sensing capabilities and be more intuitive about what to do within different contexts. This seems like a big change when you imagine it on a global scale, but some of this technology is available today, although it is not yet widespread.

More products and services will be adjusted to fit into global microcultures. Through artificial intelligence, companies will be able to quickly and easily adapt products and services to a particular country or area, minimizing risk to their brand.

Products will take ecosystems into consideration and not just be human-focused. Ecological concerns have been around for a while (such as ivory, or teak from the rainforest) but such concerns will become more widespread. Design has been human-centered for decades, and this will remain the major consideration. But with global warming and animal extinctions looming, other considerations for a product or service will have a louder voice.

Products in the future may become less gender specific. Distinctions may begin to blur for products originally designed specifically for male or female use. As cultures and lifestyles continue to change, companies may find it easier to not designate gender for certain products. In Jack Myers book *The Future of Men, Masculinity in the Twenty-First Century*, he points out that "the future roles of men and women in society, culture, and business will take decades to evolve . . . and, only when our society becomes more comfortable with current expressions of both heterosexual and LGBT lives will we learn more egalitarian ways to work together."[8]

We may have fewer objects in the future. As more hardware activities migrate to software functions, you can expect a decline in physical products. In *Design is the Problem*, author Nathan Shedroff suggests that, instead of simply designing the next great mobile phone, design solutions of tomorrow might instead explore the question of whether we need a phone at all.[9]

Problem statements for design in the future may not involve the presupposition that there will be a hardware artifact. You can look at the general need and design something that is better integrated into our ecosystems and lifestyle.

Humans have evolved alongside possessions. You use physical objects, and you have an emotional attachment to them. In *The Form of Design*, author Josiah Kahane ponders the possibility of a post-object society:

> "Yes, I accept that several highly physical families of products will completely disappear from our life, but I am certain that other objects will rise in their place. . . . We will not be left with nothing to look at; with nothing to touch, feel or smell. . . . After all, we should bear in mind that human propensity for adornment is almost as ceaseless as is our toolmaking. We will continue our affair with man-made objects, though the nature and form of these objects may change profoundly."[10]

What you think of as good design attributes may change in the future based on what becomes culturally important. You may find the future to be free of most products. If convenience and costs are critical factors in purchase, you could experience more service-oriented economies in the future, with menu options for rentals to fit your lifestyle. What will be your new leisure? What will be your new security? What will be your new wealth? You have a glimpse of some of the answers from emerging technologies and cultural trends, but cannot be certain of future values. Some attributes will be common, and people will relate to them, but others will continue to be culture-centric. The next sections will shed light on what makes for good design attributes today and how that might change in the future.

What Does This Mean for the Future of Design?

Designers will need to spend more time understanding the scope and context of what is needed for global design and what more is required of them to find effective solutions. In order to provide a meaningful emotional experience, they will need to deeply understand the culture(s) for which they are designing, and this may involve more user research at different stages in the design process. Designers can also help companies avoid risk by stopping the release of unacceptable products or services into particular cultures. This will add value to the brand in the long run.

Designers, in cooperation with marketers, will need to translate the ethics and mission of a company, which are part of the value of the brand. They will need to communicate the impact that the brand has on the world, whether it is animals, climate, ecology, humans, or all of the above. The brand's impact might also involve communicating service-oriented activities performed by employees.

As cultures shift, we will find ourselves responding to changes with sympathy or fear, delight or excitement, depending on our beliefs and the circumstances. Companies and designers will be responding to these rapid changes as well, hopefully with aid of good data, adequate information, and a great design team.

CHAPTER 5

Spaces for Innovation

In some ways, creativity can occur anywhere and does not need a special environment. People have always made new things from found objects or used whatever space was available to create: a garage, a kitchen table, or backyard. A garage or kitchen table might be the only space available, but it has another component: your comfort and familiarity with the environment. There is something wonderful about a space that allows you to relax into focused thinking about a problem and have all of your tools at hand, whether it is a yellow pad and pencil, or your old laptop computer and a cup of coffee.

A physical environment to support design and innovation at work is easy to achieve today if you have the right spaces and equipment. This chapter will review a few companies that have created great spaces to support design and innovation. They have provided comfortable, private spaces as well as larger spaces for discussions and critiques. Also available are those in-between spaces that support specific tasks, such as model making, user testing, video documentation, and the chance to talk quietly with a colleague. In the future, this list might also include virtual and physical spaces to accommodate large teams and new communication technologies.

When a team is going through the design process, it is important to have the optimum space at the worksite. I am not suggesting expensive surroundings or hiring an interior designer for a particular "look" (although you could). The creative space, whether physical or virtual (or both), enables the team more operational space for the different activities that may occur. Team discussions, brainstorming, drawing, reviewing research, presenting, documenting, and more will occur in this space. The team members also need a quiet space where they can reflect, read, and listen to their thoughts.

When teams are in different parts of the world, a virtual space is needed to bring everyone together. Working across the globe adds to the difficulty when you are working in different time zones. If the team has grown large due to the complexity of the project, this adds to the challenge as well. New research on supportive work environments from companies such as Steelcase and new generative software from companies such as Autodesk will help to create new physical and virtual spaces for innovation around the world. With the onset of big data sets to capture trends, artificial intelligence to predict best solutions, and the refinement of virtual spaces for testing, part of the workspace will change.

Some companies have these design-friendly workspaces today. But companies and countries that are new to design and innovation processes will need to invest in creative spaces. Lack of resources, physical space, or management support can keep design teams in uninspiring places with uninspired ideas. Idea generation may be hindered if employees are not empowered to engage in the needed stimuli and space to create.

As stated earlier, China has made investments in science and innovation parks and art hubs in most first-, second-, and third-tier cities. Beijing, Hong Kong, Shanghai, Shenzhen, and Wuxi all have several innovation parks and arts sections in and around their city to support new businesses and brands. This investment is a sign that they are supporting innovation on a large scale and have the resources to do so.

This brings up questions such as:

- What types of physical spaces are needed to support design in the future?
- Is a physical space even necessary, or can the act of designing exist totally online?
- What will happen to the "third space," the coffee shop, or places outside the office?

In this chapter I will look at the benefits of having a physical space optimized to support design work and design thinking. I will also consider what is needed for online and virtual work and how workplaces might change.

Old Model, Old Benefits

Having workers report to a company and sit at their desk for most of the day is a nineteenth-century model that has been slow to change. The message to the employee is, "We give you a salary, we want you here where we can find you." Older workers in the United States might be familiar with the management attitude that implied, "We own you until you retire." Things are changing in some industries, but this standard still persists—to the frustration of many employees. Today, progressive employers want their employees to think, "I like it here and I am inspired." For high-tech companies competing for the very top hire, they want to ask, "Do you like this place better than the competition?"[1]

A benefit of the old model was that you *could* find someone at their desk when you needed them. Today, of course, there are other ways of finding colleagues and communicating. As work moves into the future, physical work models may start to loosen as companies move away from lavish work lifestyles. A few major tech companies are already offering their workers money to move away from high-priced areas such as San Francisco, California, and work remotely. Matt Mullenweg, founder CEO of Automattic, the parent company of WordPress, offers stipends to employees who work from home.[2]

City Headquarters, Suburban Campus, Both or Neither

City headquarters and suburban campuses seem to be the two main types of spaces that attract companies. However, there are signs that these environments may change in the future as communication software grows, cities become costlier, and the need for physical space declines. Yet as seen in the next section, some cities are planning for the future and making updates in order to remain attractive to companies.

The Responsive City Space

Cities weren't always designed around people. They were often designed around transportation needs, whether it was horse and cart or automobile.

Many cities grew up around a castle, distribution center, or a manufacturing plant. Anne Stenros, Helsinki's first chief design officer, believes cities should be designed around people.[3] Stenros says that planning and designing cities should involve design for the "greater good," with future generations in mind. "Cities will evolve from being 'smart' to being 'responsive.' . . . The city could and should proliferate an abundance of ideas, inspiration, and creativity for all kinds of people."[4]

Responsive cities of the future can give back to their inhabitants in many ways. In the future, sensing technologies and other technologies in the infrastructure of cities that can connect to personal devices will become more common. Technologies and responsive cities can help people navigate more easily, stay safe, and become aware of events, entertainment, education opportunities, or open parking spots. Imagine you are heading into a high-crime area, and you receive an alert to avoid the area. Or maybe you receive an alert that a friend of yours is nearby. Some of these options are available today, but they are not widespread.

Alexander and Jochen Renz, managing partners of New Mobility Consulting, advise corporations, start-ups, and investors on how to thrive in this future by rethinking city design and car ownership.

> "As we explore the future of mobility, we take a holistic view of the emerging ecosystem of mobility far beyond the traditional automotive value chain. We believe that mobility as a service has profound implications on public transportation, car dealers, aftermarket service, and insurance but also real estate and other sectors. In fact, redesigning mobility and transportation will lead to a redesign of cities as we can reclaim parking space to create more livable and vibrant cities with pedestrian zones, street cafes, parks, and playgrounds.
>
> In light of connected, autonomous, shared and electric vehicles, the world of mobility and transportation is undergoing unprecedented change. Exponential technologies will enable new business models that will transform traditional industry value chains into a new mobility ecosystem. New business models emerge that will fundamentally alter how we as consumers think about consuming mobility (and car ownership)."[5]

The problem of city transportation and personal technologies coexisting will need to be solved. A city in China, tired of people looking at a screen when they are walking, has put in a special pedestrian lane for phone viewers.[6] Maybe the future will involve a personal virtual assistant telling you to look up from your phone when you reach an unsafe area. Or it can quietly tell you about restaurants, remind you about appointments, and let you know whether you should walk, call a shared vehicle, or take public transit. This would be great if you are walking alone, but it might be distracting if you are talking to someone else.

A major benefit for companies located in the city is the urban energy that fuels the competitive and creative spirit. Urban locations also benefit workers who want to use the amenities of the city (libraries, museums, storefronts, street music, etc.) to be inspired or take a break from work. There is usually a great supply of skilled workers in a city setting, especially with competitors and universities nearby. In 2015, Gensler, one of the world's leading collaborative design firms, conducted research that resulted in a report called "The Future of Workplace." This design forecast showed several trends in the debate over locating in cities or suburbs, most notably a desire for companies to locate near transit, and encouragement for biking or walking.[7] So it is not a matter of either/or, but rather what amenities and activities are available in that space.

Many companies today rent or lease space if they need to quickly add or downsize employees. Other companies flex space and accommodate changes on the fly, such as changing former conference rooms into pop-up exhibit space or setting up a maker's lab (a space to make physical products) where they used to have storage. While large cities offer the stimuli and competitive energy needed for new product development, some companies may find it beneficial to move to smaller towns, or even rural settings, depending on their product. Companies drawing inspiration from natural settings may find it necessary to be closer to outdoor environments.

Not everyone is a fan of having company headquarters in the city. In a *Fast Company* article, author Stephane Kasriel says:

> "Over the last decade, our obsession with luring large corporate offices toward increasingly dense central business districts has strained American life as we know it—particularly for the younger half of the workforce. For too many cities, being an engine of fast-paced,

high-skilled job growth has also meant higher rents, longer commutes, less savings, and fewer homeowners."[8]

Beijing, Boston, Hong Kong, London, New York, San Francisco, Seattle, Shanghai, and Vancouver all have high rents, which can cause start-ups to move to the outskirts or to less desirable parts of the city. However, these moves can still concentrate labor and jobs in a densely populated area, and singles and families may find it difficult to afford a home or support their preferred lifestyle.

Campuses

Today, some companies prefer to be outside of the city in a suburban area, where their workers can afford housing and schools. Campuses also enable companies to provide facilities that support a balanced lifestyle at work. This includes physical spaces for creative work, such as design and innovation, to take place. In the United States, Apple, Nike, Google, and other high-tech companies have demonstrated the value of great facilities for their workers and have raised the bar for many companies. Some of these campuses have daycare for children, a dog park, free lunch with healthy choices, exercise rooms and swimming pools, bike paths, and more. These perks are exceptional, but only wealthy companies can afford to provide such an optimum work environment.

The value of a campus is that it gives employees an academic feel of walking to class and seeing friends and colleagues along the way. It provides a feeling of belonging to something bigger. Apple Park, often called the "spaceship campus," is futuristic in that it has a circular design and is powered by renewable energy. It encompasses 2.8 million square feet and houses 12,000 employees. Amazon's large glass domes will have waterfalls and plant life. Google's new campus will be an 18-acre site that will include a public park and walking paths.[9]

Having the right physical space for employees, whether it is in a city, a campus, at home, or in another space, is critical to productivity. Spaces communicate a feeling. Is the space welcoming and engaging? Comfortable and easy to use? Free of distractions? Or does it have great places to go when you want to "get away for a while"?

In the next section I will review a few companies that were able to transform space for the future.

The Farm for the Future

Just outside of Venice, Italy, on a fertile plain, is h-Farm, a unique company and environment that is the brainchild of Maurizio Rossi, its cofounder and vice chairman. h-Farm was one of the first environments in the world to support start-ups, and it still engages with specially chosen technology start-ups. It offers a variety of experiences for students, entrepreneurs, and companies; at h-farm you can engage directly in design and technology. It is a school, as well as a place where companies go to understand the new technologies and how they might benefit from them and integrate them into their companies.[10]

Tomas Barazza, h-Farm's industry-human innovation culture director, helps organizations understand their digital culture and its impact on their organization.

> "Innovation can no longer be viewed as a single "light bulb" moment or the application of sophisticated technologies; it is rather about the ability to maintain a human perspective and strategically create new solutions that not only bring tangible value but also focus on people and their most intimate needs.

FIGURE 5.1   h-Farm location near Venice, Italy.

FIGURE 5.2   Experiencing technology at h-Farm.

Working to create a culture of innovation means providing a fertile ground that nourishes individual potential needed to transform an idea into reality while envisioning the future. It is about setting the scene and facilitating the mental and emotional process associated with creating something new or finding a new way of doing something old.

Today more than ever organizations have the opportunity to transform their wealth of knowledge, slide out of their comfort zone, and design in a free exploratory manner a new envisioned future free of structural constraints and technical inabilities.

It is a moment of cocreating and freedom of experimentation."[11]

FIGURE 5.3   Example of a networking and relaxing space at h-Farm.

If you are a student you can study the latest types of digital literacy. If you are a business leader, you can send your employees to train with a new technology. h-Farm is a unique physical space consisting of various types of structures and environments for training, working, gathering, presenting, dining, and enjoying life.

When I visited h-Farm for a conference, I realized that each space was purposely built to give a different feel and inspiration. There were spaces to experiment with technologies, spaces to reflect on your work, group environments where informal conversations could happen. Although everything about it pushes high-tech endeavors, the location has a homey feel. You can relax and still focus on cutting-edge opportunities.

Steelcase (USA)

For decades, Steelcase (located in Grand Rapids, Michigan) invested in research on the worker, the work environment, the education environment, ergonomics, materials, and more. Their research about how people work and the future of work has been extremely valuable in creating innovative spaces and products. Steelcase regularly shares its findings

FIGURE 5.4   Steelcase collaborative educational space.

FIGURE 5.5   Steelcase informal workspaces.

FIGURE 5.6   Steelcase combination of quiet private space and collaborative discussion space.

FIGURE 5.7   New Silq chair from Steelcase that is highly responsive to postures of the human body.

and reports, making the company an important forerunner in human-company relations.

In a 360° Steelcase Global Report, "Engagement and the Global Workplace," Steelcase partnered with the global research firm Ipsos to study 17 countries, where they interviewed 12,480 participants about their engagement at work. Key findings from this research include:

1. Employee engagement positively correlates with workplace satisfaction.
2. Engaged employees have more control over their experiences at work.
3. Fixed technology exceeds mobile technology.
4. Traditional workstyles persist.
5. Cultural context influences engagement levels.[12]

Jim Keane, Steelcase CEO, said of the report: "Its key findings affirm our belief that the places where people work can influence not only productivity, but also shape employee attitudes and beliefs."[13]

Steelcase brought professional and affordable design to the workplace in Asia. The company has improved an entire generation of work environments in China, Hong Kong, and other areas of Asia that were noticeably in need. By designing comfortable, ergonomic, and interchangeable pieces of work furniture, Steelcase was able to introduce to a new class of companies the importance of good work environments.

Steelcase also engaged in research into different types of spaces employees might need throughout the day to be most productive. Some employees were content to work alone in their offices when they needed to concentrate. At other times they needed a space like a lounge where they could meet other employees for an informal discussion. Private space was needed for team discussions and working sessions. These newly designed spaces have accommodated the growing demands for versatile, flexible, and productivity-oriented office space to accommodate people and technology.

Hong Kong

Before she died in 2016, the architect Zaha Hadid saw the completion of her Hong Kong Jockey Club Innovation Tower on the campus of the Hong Kong Polytechnic University (PolyU). Although the campus is known for its red tiled rectangular buildings, Zaha's white curved tower stands out amongst them.

The purpose for the building was to unify the School of Design, which had been housed in different parts of the PolyU campus, into a space that would support creative thinking. Victor Lo, council chairman of PolyU at that time, felt the students needed a more inspiring creative space for their classrooms, labs, and exhibitions. In addition, the students needed space to create and store projects that were too large to store in small Hong Kong flats.

Besides the mix of classrooms, exhibition, and office spaces, Hadid designed "third spaces" throughout the building, where students and visitors could go for coffee or just sit comfortably for informal conversation. The building's monumental size and unique shape seem to say "be bold" as an inspiration to students.

Aspects of Physical and Virtual Space

Responsive cities, flexible company headquarters, and third spaces all need to address the needs of the people who will work there. More specifically, the spaces for work and play need to be welcoming, comfortable, inspiring, and connected. The work environment should support different types of activities, such as private work, team interaction, informal discussions, and chance encounters.

Design and innovation spaces need to support particular activities. The team will hold regular review meetings, plus brainstorming or creative prototyping sessions. In addition, the spaces should accommodate drawing, role-playing, and informal user research. This may not all take place in the same room, but the company space should support these activities. The team also will need space for gathering and displaying visual information, and may require space to create physical models.

Specific technologies in the design and innovation spaces can help document ideas and record conversations. Quality screens with high visual resolution, speakers with high sound quality, and high data transfer speeds for large images and videos will all be a part of what is needed in the future. Versions of these technologies and products are available today, but they are constantly improving.

Future Spaces

Physical spaces may shift toward virtual places in time. Will designers physically enter into a virtual space with projected images onto plain surfaces? Will they use headsets and glasses to access virtual aspects of work? Will they use new software that allows them to generate thousands of possible solutions to review? Will those solutions be presented visually, verbally, or both? All of these are possible scenarios as technology is refined and tested. There will be corporate benefits to visualizing our data and ideas, but it is not yet known what type of mental or physical impact it will have on employees.

Future spaces will seek to provide workers with more balanced work-life activities. You can take a break with a guided meditation to relax, or play a game with a colleague for a timed interval. You might be able to work flexible hours throughout the day, but even this may entail tradeoffs. For example, would your activities be tracked and recorded so much that privacy is lost or reduced?

The physical and mental health of workers may become much more important to companies. Companies and employees may need to work together to adjust work, leisure, and service activities. This benefit could improve employee loyalty, especially for those companies that need long-term employees for complex projects.

In the future, programs and sensing technologies may track problem issues. You may see more visual recognition systems to protect workers in unsafe environments. An example of this might be an alert that you receive when there is a problem in another part of the company, giving you time to physically remove yourself from potential harm. You may also see more personal sensing technologies tracking your health during work hours. Wearing a simple blood pressure monitor can indicate exactly how annoyed you are with a work colleague!

What Does This Mean for the Future of Design?

Companies and organizations that use the design process will understand the need for creative physical and virtual spaces to support innovation activities. This involves a steep initial investment in resources and time, but having these spaces available will increase team productivity and serve as a vital foundation for coming up with new solutions and opportunities.

Future design work may not be conducted in physical facilities at all, but instead created and tested online. Designers will need to help create these virtual processes and spaces that they will use in the future. It cannot be left to technologists and managers to make such decisions. An example of this would be a company that wants designers to enter a virtual environment to solve problems. Designers should engage in that process in order to help design their new tools.

Merely having a physical creative space or having the design team virtually connected is not enough. In the next chapter I explore the importance of managing the future design team. Getting a team together physically may not always be possible, but a good team manager will keep the members engaged and on track whether they are physically together or scattered across the globe.

CHAPTER 6

Supporting Design Teams

In this chapter I discuss issues related to supporting design and innovation teams in the company, at home, and abroad. Everything from one-person companies to mega-conglomerates will need a team to complete most any project. The good news is that in the future it will be easier to find design team members from around the world as the numbers of designers increase.

Today's teams can encompass different disciplines, different experts—whomever will add value to the end result. Global teams can be structured in endless ways, and methods can vary by company. As design problems become more complex, design teams may need sub-teams to help with the increase in design tasks.

A basis for a rich culture of design and innovation must come from company leadership and be reinforced by management. First, I will discuss general points about corporations, teams, and cultures before moving on to a discussion on global teamwork.

Corporate Culture

Companies that want to take leaps in innovation and revenue must provide an effective support system. The overall company culture will ultimately drive innovation and support the design process of the future. Leadership is needed more than management in order to have a supportive culture for design.

In the book *Leading Change*, author John Kotter, writes:

> "Even today, the best-performing firms I know that operate in highly competitive industries have executives who spend most of their time leading, not managing, and employees who are empowered with the authority to manage their work groups."[1]

He believes this trend in leadership will continue despite employees and managers who want to keep the older models of management in place. However, management in some areas of the company will still be necessary. He continues:

> "Some business futurists write as if management as we know it will disappear in the twenty-first century. Everyone of importance will become visionary and inspiring. Those boring people who worry about whether inventories are on target will no longer be needed. But this is unrealistic."[2]

Companies will still need people to keep their daily activities on track. Even though some authority and responsibility will be pushed further down into certain teams, not everyone will be a leader. However, it will take leadership, not just management, to support creative teams and drive innovation, especially since design and innovation involve change. Leading a creative team and the innovative process means understanding the design process and supporting it, rather than just managing it by checking off progress made.

Being a Designer or Innovator in a Company

Colliding corporate cultures, uncooperative team members, and unsupportive leadership can impede the success of a team, the growth of a company, the growth of employees, and the job of the design manager. Being creative in a company can feel like a lonely job when designers recommend change that others might not like. In addition, colleagues and managers in other silos can develop negative feelings toward designers—or more specifically, their work—because of fears that their position will be adversely affected by a project.

Nicholas Partridge, senior innovation director of LPK in Cincinnati, Ohio, conducted a mix of qualitative and quantitative research on innovators within corporations. For him, the research confirmed "the fact that many corporate innovators, while they might have a team, can feel like lone wolves. They are seen as the outsiders inside of their companies. There's a core business that's humming along and sometimes they are regarded as almost a distraction, an annoyance, because innovators are naturally the people who are challenging the status quo."[3]

It is often difficult to be a creative person in a company where change is often viewed as unnecessary and difficult. It may be hard to get buy-in from colleagues, and you may experience outright resistance. At any level, it can be a lonely work experience to promote change when you know there is anxiety and opposition to new proposals.

Techtronic Industries (TTI), Hong Kong, is a successful company that manufactures a variety of tools for markets around the world, and has a robust design team. Alex Chunn, vice president of innovation and product planning for TTI, said key practices help his company effectively manage global planning, research, and execution of innovative and impactful product solutions.

> "Firstly, industrial design had a seat at the table when it came to planning product roadmaps and relevant user and technology research. If nothing else, this helped to ensure that there was a user-centric focus at all stages of planning and development.
>
> Secondly, a reasonable amount of budget autonomy is important. This is critical in the early phases of the project where activities can be quite divergent, iterative in nature, and difficult for others to sometimes understand the value of these early activities."[4]

Because TTI saw the value of conducting research in various stages of the design process, the company was able to develop highly successful designs internally.

Managing and Leading Teamwork

Every team is different, but there are basics to good teamwork. In an *Inc.* magazine article, Justin Bariso reported that the most important aspect of an effective team was "trust."[5] Who was on the team did not seem to matter as much as how those members interacted with each other. This was described by Google in the following way:

> "In a team with high psychological safety, teammates feel safe to take risks around their team members. They feel confident that no one on the team will embarrass or punish anyone else for admitting a mistake, asking a question, or offering a new idea."[6]

It is imperative for leadership to make certain that teams do not fall into patterns of homogeneity, such as always establishing teams that are young, all-white male engineers, as has happened in many technology firms. When teams are nonhomogenous, however, it may take longer for trust to develop within the team.

While there are many good books and articles about managing and supporting teams, the research of Martine Haas and Mark Mortensen provides an insightful and useful framework. In their *Harvard Business Review* article "The Secrets of Great Teamwork," they provide solutions for teams today, which are more "diverse, dispersed, digital, and dynamic (with frequent changes in membership)."[7] From research on teams in a variety of settings, Hass and Mortensen describe the right conditions for teams:

- **A compelling direction**, having a common goal that is not too difficult, but challenging and clear
- **A strong structure**, having the right mix and number of members
- **A supportive context**, provided with the right training, resources, and information
- **A shared mindset**, where members freely share information and have a strong common identity[8]

These suggestions, along with trust, form the base of what is needed for cohesive and effective teams today.

Design and Innovation Teams

Mark Dziersk, executive editor of the Industrial Designer's Society of America *Innovation* magazine, wants to put to rest the idea that innovation is best as a lone activity. "The creative genius, the mad scientist, the inspired rogue designer—these stereotypes mask the truth that innovation is a team sport. And when teams are properly enabled, they produce better results."[9]

Some designers may want to work alone part of the time. This is fine if it helps them concentrate or complete a particular aspect of their work. It is best, however, if creative employees can work together as a team for most of the process. They can learn from each other and—particularly in the right setting—inspire more ideas.

Although many designers and creative professionals may feel alone in a company, they need to work in teams, especially if they are working to solve complex problems. Fortunately, ongoing research of design teams specifically (and creative teams in general) helps designers to understand when a team may need to shift into other modes of operation.

Keith Sawyer, an expert on creativity and groups and author of *Group Genius*, said this about studying improvisational teams in music and comedy:

> "Improvisational groups are self-managing: They have a magical and mysterious ability to restructure and regroup in response to unexpected events without being directed by a leader. Self-managing groups are particularly effective at innovation in rapidly changing environments. The paradox of innovation is that organizations emphasize order and control, and yet improvisation seems to be uncontrollable."[10]

It is this uncontrollable aspect of where the work can go that can make design managers frustrated. However, design leaders are able to be supportive and inspire the team to great results. Creative teams may be hard to lead and manage if you don't understand the design process.

What Is Considered a Global Design Team?

Provided below are common examples of global design teams:

- Employees in the *same company* working in different parts of the world on design, products, and services. An example of this would be a company that sets up teams in different parts of the world to code, design, develop, manufacture, or conduct research.

- Employees in *different companies* in different parts of the world working on the same products and services. An example of this would be a company that hires a design or engineering consulting group in a particular region of the world.

Creative Teams Plus Cultures

There are several cultures that design managers and teams need to understand and work across. These include:

1. Their company culture
2. Their team culture
3. The culture of the country or region

All of these cultures are present for the team at the same time. Being aware of cultural differences might help sort out and avoid difficulties or issues down the road.

Roles and Decision-Making in Other Countries.

In "Being the Boss in Brussels, Boston, and Beijing" by Erin Meyer, the author notes that global team "approaches to authority and decision-making are not the only ways in which cultures differ, but they are arguably the most important in the leadership context."[11]

She cites the different attitudes toward decision-making in countries such as India, Italy, Mexico, Russia, Japan, Germany, and the United States. In some countries, decisions are made quickly but are subject to change as new information becomes available. In other countries, you might have a lot of people involved in decision-making, but once a decision is made it is considered a commitment, and the decision will rarely change. In the United States, consensus decision-making has historically not been the norm in business, and so decision-making frequently lies with the boss.[12]

A *Business Insider* article titled "24 Charts of Leadership Styles around the World" cited the work of Richard D. Lewis, pioneer in intercultural communication, who provides insights into generalized management styles, with findings such as:

- "British managers are diplomatic, casual, helpful, willing to compromise."[13]

- "In Latin and Arab countries, authority is concentrated in the chief executive."[14]
- "American managers are assertive, aggressive, goal and action oriented."[15]
- "French managers tend to be autocratic and paternalistic."[16]

Lewis cautions that you cannot fall into cultural stereotypes when assessing leadership styles, but there are some norms within each of the countries.[17]

Eastern and Western methods of creativity are different from one another and effective in their own way. Often, the West's "hero" approach to creativity and innovation is very individualistic, while the East's approach is team-generated and geared toward incremental innovations.

Global Teamwork

Leading design and development teams has its challenges, especially when the teams exist across different countries and cultures. Cultural differences and communication become even more complex when the teams encompass different companies and work on a 24-hour schedule.

Setting up teams in other parts of the world is attractive to companies because of the many benefits.[18] Companies benefit from setting up design and development teams in foreign countries for specific expertise, lower cost, or to be closer to the culture in which the product or service will be sold.[19]

Leaders and managers with diverse, global teams will need to keep them engaged and interested in the problems to be solved, discuss the way the teams can best work together in a structure, provide support to all of their members, and strive to have a common and cohesive identity.

Regional and Country Culture Differences Affecting Design Teams

Qifeng Yan, founder and CEO of Loobot Tech. Co. in Shenzhen, China, worked at Nokia in Helsinki, Finland, for more than 10 years before returning to China to head his own service robot company and teach design and innovation at local universities. His global experience in user research has given him a world of experience with global products. I asked Qifeng (pronounced Chee-fung) if he thought design thinking was affected by culture.

> "Design research, because of cultural differences, may need a different methodology. For Eastern people such as [the] Japanese, it can be very easy to form a collective or group idea; their way of thinking is to answer one question all the same in order to influence each other. In Helsinki and the United States, you don't find this situation. They are more difficult to be influenced by each other as individuals.
>
> In China, if you discuss in a group, you will have one design at the end. We have a national collective culture, with one thought in the end, although designers in China now work in a more solitary way in the beginning of the project and will have ideas individually. Only after they have a design drawing will they discuss it. Designers in Europe brainstorm together in the beginning, then they have their own thoughts after the discussion. The cultures are different.
>
> Chinese people are very accepting of new things and care more about convenience, and care less about privacy than Western people. Much of this convenience is about location. Mobile payment in China allows sharing bicycles, power banks for electronic recharging, and even umbrellas. Use your mobile phone to pay the deficit, use the product, and return it back."[20]

When I asked Alex Chunn of TTI what some of the difficulties were with design for global products, he said:

> "The challenge was always understanding the key markets and how each had their own market forces at play. For example, North America has strong retail ties and we need to understand the retailer's strategy and objectives. The target users tend to be more homogeneous in nature, whereas the European market is more fragmented with different key retailers by regions, such as Nordic, Mediterranean, and the United Kingdom, and [they] tend to be more heterogeneous in comparison to what we would see in North America.
>
> Given these challenges in a fast-paced, globally connected product development environment, it is important that designers 'know their stuff' and can clearly define value propositions based on user insights."[21]

Issues such as work hours, leadership style, and compensation vary widely from country to country. These differences add a layer of complexity and information for the design leader. New information from the Organization for Economic Cooperation and Development (OECD) report lists Germans as working the fewest number of hours in the world but they are also extremely productive.[22]

As the number of global teams increases in the future, we will need to have a much greater understanding of working with and in other cultures. We need to understand the interplay of global cultures and workgroup dynamics. Hopefully the global design manager who has teams in different parts of the world will understand the nuances of the design process as well.

Sending Knowledgeable Team Members Overseas

At some point, team members may need to go to other countries and regions. It is more important than ever that they gain cultural awareness in context of how they are perceived and how they perceive others.

Learning about different cultures is not just about business etiquette; you also learn what *not* to say. Discussing certain political issues or criticizing rulers, religions, and/or cultures is not always a good idea when you are in another country. For example, in Thailand, criticizing the ruler is punishable by law. Products that deride political leaders would be very unwelcome if you were in China, Russia, or other countries with powerful figureheads.

The cultural differences can be extreme. The constant focus on Western individuality can make Easterners uncomfortable. Western speech is often spoken from the first-person "I" point of view, and can seem very self-centered, domineering, and insensitive to the concerns or issues of others. In addition, Westerners' openness to discuss almost anything can put global colleagues in an uneasy position, especially if it is a leader-subordinate relationship.

Eastern sensitivities, however, can be lost on Westerners. After living in Asia for seven years, I started to marvel at how individualism took a back seat. Even the speech patterns were different, with conversations rarely starting out with "I." Humbleness is part of the Chinese culture, but it does not mean they are not proud of their accomplishments. They just do not advertise them as much as people do in the West.

Even polite Chinese gestures such as leaving a bit of food on plates at the end of the meal are at odds with the "clean plate" culture of the West. Empty plates in China at the end of a meal might mean that the client or company did not prepare enough food for the guests, whereas in the West, eating everything on your plate is high praise for those who cooked and served.

All of these cultural differences pervade global work teams. It is important, however, to be respectful of those cultures, even if their norms are not integrated into the teams. Cultural awareness and sensitivity training need to be a constant part of the future.

Design and Innovation Team Communication

Today, managing a global team for product innovation has, in some ways, gotten easier. Long distances are bridged with a text, email, phone call, video conference call, and other types of communication and collaborative software. Productivity has risen for companies that can keep work going around the clock, around the world. What has *not* gotten easier is the need to keep people in different parts of the world feeling like they are an important part of the same team.

James Ludwig, vice president of global design and product engineering at Steelcase, has teams in Hong Kong and Dongguan, China; Munich, Germany; Grand Rapids, Michigan (USA); and Kuala Lumpur, Malaysia. I asked him how he manages to keep his teams working well together when they are in different parts of the world. He said:

> "I try to foster and support good working relationships *between* the teams. If one team is trying something and would like feedback, they will often work together through the team leaders before I am approached, and I will always ask if other teams have been consulted. We believe the ideas get better through this process of peer review and dialogue."[23]

When I asked James about future technologies related to the workplace, he said the work will be both "easier and difficult." Easier in that the technology will allow for quicker iteration, process changes, and simultaneous collaboration remotely at a blistering pace. However, an appetite for change (and an ever-faster product cycle) can be the challenging part. This is largely due to the same forces that allow designers to create more rapidly.

As for meeting with the teams, he said, "I am still biased toward the smaller, personal meetings, and travel to the sites frequently, although larger meetings are good for strategy rollouts and announcements." He is looking forward to new and refined communications technologies that continue to erode the presence disparity.[24]

Managing several teams in different places requires keeping everyone informed, otherwise it is like "handing off" the work, leaving the receiving team without any mental preparation for the project. Alex Chunn of TTI, said:

> "We worked hard to maintain ties with our fellow designers in other regions and business units, and so we had strength in numbers. These relationships were important to help us with training and [with] understanding user and market dynamics that might not be captured in a product brief, helping us with the overall management of the user experience, and with the visual aspects of the brand identity."[25]

What Does This Mean for Corporate Leaders in the Future?

If you are not already becoming more supportive of design and innovation, the time to start is now.

Proctor & Gamble serves as an excellent example of a large company that incorporated major changes to its management style years ago and is reaping rewards today. In a *Harvard Business Review* article called "How P&G Tripled Its Innovation Success Rate," authors Bruce Brown and Scott D. Anthony explain how the company created a "new-growth factory" that helped P&G change its corporate culture to focus on innovation.[26] P&G started transforming the company through their "factory" as early as 2004. This "factory" was made up of activities that:

- Teach senior management and project team members the mindsets and behaviors that foster disruptive growth
- Form a group of new-growth business guides to help teams working on disruptive projects

- Develop organizational structures to drive new growth
- Produce a process manual—a step-by-step guide to creating new-growth businesses
- Run demonstration projects to showcase the emerging factory's work[27]

It took years for P&G to change its culture, but the change brought the company a successful future.

Looking forward 10, 20, or 50 years, you may have to imagine what the best culture will be for your company, and how you can manage that process of change now. What will companies be grappling with in the future? One thing that will not change for companies is the importance of the customer and user.

Katherine Bennett, professor at ArtCenter College of Design in Pasadena, California, discusses the success of Xiaomi, the Chinese high-growth tech-based enterprise that took on the other tech giants. She said:

> "Xiaomi management's strength is a deep understanding of the customer, and the design mindset is key to that understanding. This was baked into the business model from the beginning, according to one of the founders, Liu De, an industrial designer. 'We wanted to build a breakthrough business model that is superior on multiple levels, based on an understanding of the market and current trends,' he recalled. 'It wasn't enough to have a great product; we had to innovate in different areas of the company, and those innovations needed to be rooted in our understanding of new generation of young Chinese.'
>
> This approach is embedded throughout the company and can be brought to bear in all sorts of decisions. For example, Xiaomi partners with over 200 companies to manufacture their IoT products. Small groups of engineers are charged with making these investments and are given a great deal of autonomy, investing up to $4M without board or C-level approval. In other companies, these teams are usually made of 'numbers people.' Xiaomi's team bases investment decisions not on financial metrics but on their knowledge of the customer and the needs of the Xiaomi IoT ecosystem. Embedding design principles deeply into a company makes the enterprise nimble and provides a powerful competitive edge."[28]

The design process can help companies by allowing their employees to have an open mind toward new ideas and concepts. This is a major switch from the "We don't do it that way here" attitudes of businesses from prior decades.

What Does This Mean for the Future of Design?

The role of designer in the future could become similar to that of a curator, conductor, film director, or operations specialist, depending on the product or service. Interior designers, product designers, and user interface designers have skills today that are directly transferable to building virtual environments.

In Rob Girling's article, "AI and the future of design: What will the designer of 2025 look like?" he says that "designers may well provide the missing link between AI and humanity."[29] Designers may go into various specializations, such as virtual environments. He says that "the designer's role will be to set the goals, parameters, and constraints, and then review and fine-tune the AI-generated designs."[30] Girling does not believe that design jobs will go away, but they may change. There will be an opportunity to design AI environments and user interactions, but first designers need to help create the emerging AI tools and systems they will be using.

In the future, there will likely be a melding of work methods as cultures around the world adopt and adapt team management practices. Accepting and embracing continual change may become the most valued skill of all.

Evolving Future Roles in Design

Will design management remain familiar in the future? Or will there be teams that are self-managed with the aid of a benevolent "augmented manager"? I can only speculate, but someone (or something) is needed to ensure the appropriate technologies are employed in the design process, to verify data and research, and to identify ethical, legal, and safety issues. The roles of design team members may change according to the needs of the project. Design managers may become conductors, coordinators, curators, and staging experience experts.

Designers today can get a head start on the future by seeking out companies that are already involved in artificial intelligence. In an article titled "7 surprising companies where you can work on cutting-edge AI technology," author Ben Dickson lists companies that are using artificial intelligence within a product specialty. He listed Volkswagen for self-driving cars, JPMorgan Chase for banking, Philips for healthcare, Panasonic for computer vision, Palo Alto Networks for cyber security, Affectiva for emotion detection, and Deeplearning.ai for training.[31] Of course, these are not the only companies where designers of the future can engage in evolving their work and improving their skills, but it can be a starter list for future opportunities.

Designers and entrepreneurs working on their own may find that technology will help bring about easier design solutions. There will be a steeper learning curve for engaging with the new technologies. The designers working alone will find it easier to tap into the expertise of others and find work partners across the globe, if needed. They will find that learning the culture of their partners can be a good investment in their time, especially for projects conducted with that culture.

What Does This Mean for Global Design Leaders in the Future?

The following issues for global teams and team management are paramount:

1. Supportive leaders who enforce a company culture that embraces design and innovation. This takes years to perfect for an older company, but even younger companies can be modeled on older, outdated practices. This support for creativity and innovation should come in the form of resources, environments, and attention to creative processes within the company.
2. Communication among global teams, headquarters, and each other on a regular basis. Communication can come in traditional forms today, but in the future, you may be getting updates from "teammates" who are not human, but rather supportive artificial intelligence systems.
3. Establishing trust and a unified mindset among the team members to help them feel cohesive and that they are all in the "innovation space" together, even if they are on different continents. In the future as technologies improve, it will be much easier for team members to be in the same space, even if it is virtual.

4. Education and sensitivity training for global work cultures and global cultures in general. Any global endeavor will require information on the targeted cultures in order to reduce risk for the company.

Senior managers take risks when they move decision-making, but there's already a shift under way of companies moving decision-making authority into some nontraditional areas. In the future, however, much of this risk-taking may be aided by artificial intelligence programs collecting large amounts of data and sorting out trends. That is the promise of using technology in the future, and some of these tools are already here, just not refined or robust enough yet to be useful to many for particular applications.

CHAPTER 7

How to Evaluate Product Concepts

Safety is the most important aspect of a product or service. Products should not bring harm to people, a company, or the environment. The goal of evaluating concepts is twofold: reduce the risk of harm, and design the best product possible.

This chapter will present information on making evaluations during the design process. These evaluations are offered to help ensure low-risk design and innovation. This chapter introduces examples of design thinking and reasoning to help assess whether an idea should be dropped or developed.

If we follow the design process and evaluate products and services at the right time, we can avoid risk and other problems. It is far better to be aware of issues or problems *before* investing in a product or service, and the design process is set up with evaluative checkpoints throughout.

While best practices for evaluating products and services are presented here, I will also consider how we might evaluate products in the future. There will surely be new ways to test products and services in software or in virtual worlds. There may also be a need to broaden and diversify research methods to adapt to other cultures. Western research methods are not universally accepted in some cultures, and it is important to adapt or invent new ways of obtaining information.

The design process can have many methods of evaluation and research, which can be conducted throughout all three stages. At times, designers

TABLE 7.1   The Design Process Evaluation Checkpoints

Stage I *Problem or* *opportunity inquiry* →	*Stage II* *Solution* *generation* →	*Stage III* *Solution* *evaluation*
Problem statement origination	Idea selection	User testing
Big data	Innovation difficulty	Cultural issues
10 impact factors	Sketch prototypes	Consumer research
Research on country/ culture	Critique	Reliability testing
Early user testing	User feedback	Sustainability evaluation

may only need one or two methods to get valuable feedback. Table 7.1 lists evaluations that can be done to help the team reach optimal solutions. Some of these methods are traditional research methods, while others are discussions that should occur during the design process.

Evaluation in all Three Stages

Stage I: Problem or Opportunity Inquiry

In stage I, information is gathered from a variety of sources: potential customers and users, new technologies or materials, competitors' products or services, cultural trends, and more. Specific information about a country or region may be needed, along with other information pertinent to the challenge.

The evaluation and research during this first phase vary. Interviews, focus groups, or surveys might be conducted to gain general knowledge, or materials (photos, drawings, patents, etc.) might be collected and posted for viewing. In other words, you are collecting information that may inspire the team. In this stage, the problem statement is reviewed for clarity, level of complexity, or other issues that might pose obstacles to new designs or innovative ideas. Evaluation can occur throughout the design process, starting from a diverging or opening viewpoint to a convergent or narrowing viewpoint.

At the beginning of the process, you *evaluate for opportunity*. You constantly search for pieces of information that might work together. What new materials or technologies can be used? What would be considered innovative and unique? Helpful and sustainable? What would support the brand and the workers? You are seeking opportunities at this stage and do not want to rule anything out. As you start to form viable ideas, you need to evaluate them. This leads to stage II, where the process flips from *evaluating* to *eliminating* ideas in order to focus on the best ones. Mockups and prototypes are made to test the ideas.

All products and services should be evaluated against their original problem statement. New information from user research or other sources can greatly impact the original problem statement. If a major flaw is found in the proposed solution, and the proposal is a direct result of the problem statement, senior management should be informed so that the problem statement could be revised. For example, a problem statement might ask for designs for new motor scooters in a particular country at a particular price. Halfway into the project, the targeted country may issue an edict that only electric motor scooters can be used. The company making the scooters will have to look at another country in which to sell their scooters, or alter their original statement to include the electric requirement.

Problem Statement Origination

The team should read and discuss the problem statement together. They should ask questions, such as:

- Is this a clear statement?
- Is it too open-ended?
- Do we need to narrow the scope of the problem?
- Is the statement too defined?
- Are we missing out on other potentially revolutionary ideas?

Here is an example of a problem statement that may be too broad: *Design bathing products for people with disabilities*. This statement may lead to a lack of focus; there are too many disabilities to consider at once. Open-ended problem statements can lead to innovations, but they might not help the company if the new ideas are not something they can make, market, or distribute. Here is the other side, an example of a problem statement that may be too narrow: *Design a soap holder for elderly women.* This statement will leave out other parts of the population who may need bathing products. A problem statement that is "just right" is one that provides room for new ideas and innovations but is not so wide that it lacks focus. Here is an example: *Design a new bath seat for seniors that will also accommodate their bathing supplies for easy access.* This team could come up with a great idea for a new bathing seat to be used in a bathtub or shower, but one that other people will find useful, too.

Big Data

Design teams often turn to data in relation to the problem they want to solve. It can be consumer data, technical data, or even a visual representation of all the competitor's products. This data needs to be vetted for accuracy and applicability to the problem, and the teams need to know how to use and vet this information. The team should have an expert in data analysis, if possible, and at the very least the designer should have a working knowledge of research and statistics. The team should not be swayed by statistics if it does not address the problem, or if something seems amiss. Designers in the future will need to study statistical research in order to understand where it is useful and where it may sway an idea; such data can greatly affect the design reasoning for why a solution was chosen.

10 Impact Factors

As discussed earlier in this book, the impact factors will give you a broad picture of the issues you need to consider when evaluating your idea. You can prioritize the burning issues that need immediate attention. Will you be allowed to sell in that country? What are the requirements? Will your country allow you to export? Those are some of the important questions that need quick answers—and which may require additional people on your team to troubleshoot. Is there a sustainability issue? Find someone who deals with international material cradle-to-grave issues. Is there a potential safety issue with the product? Seek international legal advice.

For example, consider a company that wants to sell its products in South America. To gain basic information, the team should study the history, cultural trends, political movements, technological infrastructure, and natural disaster propensities. This data can help make decisions easier about which country to enter first, or whether to enter the market at all. A company might instead decide to look for opportunities elsewhere, such as Africa.

Research on Country and Culture

Research on country and culture is critical for gaining knowledge that might affect the outcome of your product or service being offered. A historical search of the region is recommended as well to look for patterns of war or peace, frequent changing governments, changes in laws, and more. This robust information will help the company decide the level of risk involved in launching the product or service.

Stage II: Solution Generation

Idea Selection

If you had a representative photo of this point in the design process, it would show a team looking at walls covered with Post-it notes. Concept evaluation can be one of the most difficult parts of the process because you need to make sense of all the information and come up with solutions.

In *Change by Design*, author Tim Brown succinctly captures this point in the process. "Synthesis, the act of extracting meaningful patterns from masses of raw information, is a fundamentally creative act; the data are just that—data—and the facts *never* speak for themselves."[1]

At this point, a team can feel overwhelmed with information, or underwhelmed if the ideas are mediocre. This is where courage enters the picture.

Innovation Difficulty

At this point the team might feel like a great new idea or innovation just isn't possible, so they might return to the problem statement and review it for inspiration. Often, specific conversations, biases, or other issues arise in the early stages, and this sets the team off in a particular direction. The team may need to go back to see where a particular line of reasoning branched off, giving them a chance to choose another path. For example, imagine that a team is tasked with designing rescue equipment for flooding disasters, and the team moves in the direction of an inflatable device that would transport a family to safety. The team might have moved in this direction based on a new material that withstands punctures while inflated. In hindsight, the solution may not involve a rescue vessel at all, but instead may involve in-home alerts with specific directions for evacuation.

Sketch Prototypes

Often something is needed to look at or point to when discussing a product or service, so it is important to visualize ideas. This visualization can be in sketch form (a quick drawing), rough 3-D form (foam shapes or cardboard mock-ups), or in storyboard or simulated computer interfaces. The value of visualizing ideas is that everyone on the team can interact with and further define the product or service. Visualization also is helpful for informal user testing.

A team may put several 3-D shapes for a new tool in front of potential customers and ask them which shape most appeals to them or works best for them. This gives the users something to respond to. Users might respond with comments such as, "It looks as though this will hold up under pressure," or, "That hand grip feels too small." Testing for a service or computer software program can be as simple as showing potential users a scenario made from storyboards or a sequence of actions mocked-up on the computer. Even a simple visualization can help participants evaluate the user interaction and experience.

Critique

Critiques can happen anytime throughout the design process and they are critical when solutions need to be evaluated. Critiques are usually done as a group, during which team members can explain proposed solutions or explain why they might not like a particular idea. During a critique, some team members synthesize and filter the comments to inspire new or better ideas. They are evaluating the information emotionally and intellectually. Emotionally, one might say, "I think that solution is just fabulous." Intellectually, one might say, "This particular solution would address the sustainability issues that engineering requires."

In this phase, ideas are tested to see if they can move to stage III, Solution Evaluation. Stage II might have early user testing with rough prototypes and simulated mock-ups to catch early problems in the design. At this critical point, the design team has options:

1. Go back to re-evaluate the problem statement
2. Go back through the process to generate more ideas
3. Move forward with what you have

At this point, either an individual or a team must have the courage to speak up and say, "I think we should go with these top three ideas," or, "I don't see anything here that excites me or solves the problem." This is a very simplified view of the process, but it captures the basic activities.

Stage III: Solution Evaluation

Once you have landed on that great idea, you need to conduct more research and evaluation with refined prototypes. The prototypes at this stage should be as close to the end result as possible. This means they must be realistic-looking products or functioning user experiences in order to elicit more exacting feedback.

User Testing

User testing for final solutions comes in many forms. It can be a customer experience audit, eye tracking, participatory action research, think aloud protocol (where a user says what their actions are and what they are thinking), or a host of other types of testing.[2] As with critiques, it is important take notes when suggestions are made to improve the product. If it is your design, you may want to have someone else run the critique so you are not tempted to defend the design.

Cultural Issues

When conducting research, it is important to test for more than functionality and appeal. A product may be attractive or work well, but that does not mean it is culturally acceptable. Is there cultural bias in the product or service that we can't see? Embedding this type of evaluation in the user testing can help uncover hidden cultural issues.

Consumer Research

Traditional forms of consumer research, such as focus groups or interviews, can give us great information when the participants are articulate and knowledgeable. Unfortunately, it can be difficult to find the right sampling of people, and the information or data can be unreliable or inaccurate. When consumer research methods are correctly structured and the right participants are chosen, you can get vital information that will make or break your design. It takes courage to listen to possible flaws or shortcomings and adjust the design.

You may need to consider, however, that the participants are not knowledgeable, or can't tell you what you need to know. Some of them may be too kind to disclose a criticism, especially if they know you were involved in the creation. It is best to have others test your designs. A note of caution, however: we cannot always take methods of research and inquiry and expect them to work across all cultures. You may need to adjust your questions to fit specific cultures.

Reliability Testing

Engineers, computer programmers, psychologists, and other scientists who may be involved in the project will need time to test materials, coding, emotional effect, or whatever else might be involved. Designers need to know what expertise is needed and when—and then use that expertise in the concept-generation phases for decision-making. They will also need to document the inclusion of this expertise in the process and how it helped to shape decision-making.

Sustainability Evaluation

Sustainability is such an important issue in many countries that it deserves its own evaluation. Hopefully, the end result is that the product or service passes the most stringent sustainability requirements for any country. You do not want to start manufacturing anything without such an evaluation.

Piggybacking on that idea, it is also smart to see if the product can be repurposed. Nabil Nasr, associate provost and founding director of the Golisano Institute for Sustainability at the Rochester Institute of Technology (RIT), has helped many companies and organizations examine their products for recycling and repurposing through the Center for Remanufacturing and Resource Recovery.[3] The institute engages in the "development, testing, and deployment of efficient, environmentally-friendly and cost-effective remanufacturing processes" for products from airplanes to office equipment.[4] Understanding how to repurpose products, or recycle parts of products, is valuable information early in the design process. Design teams looking to reduce the numbers of parts in the products they are designing, or to design for reuse of specific product parts, can include a repurposing requirement in the problem statement.

Lowering Risk with Research

At the end of the Solution Generation stage (II) and beginning of the Solution Evaluation stage (III) is the point where many companies (and entrepreneurs) find themselves in a high-risk situation. They risk losing resources, time, or reputation—and could even face legal action—if they:

- Move forward with a mediocre idea, hoping it will get better as it is tested
- Don't have enough information to know whether the product or service will be successful
- Are in a hurry to get a product on the market without proper evaluations

Again, the team and company have the design process and methodology to fall back on to properly evaluate an idea. Testing prototypes with target customers before moving forward with a product helps greatly reduce risk.

It is also at this stage (stage III) that egos may enter the conversation. If senior managers are present, they may have a strong preference that can be off target because they probably do not represent the market. If senior managers are savvy, however, they will detect issues with a new product or service and help to reframe the problem (stage I) or suggest more idea generation (stage II). The team can always remind senior managers that the design team needs to go through the evaluation stage and gather more data. At best, senior management should participate or review all stages of the design process before the team reaches the end solution.

Evaluating the Design Team Rationale

Members of the design team can become very passionate about their idea(s), some of which may be tangential to the goal. That is why it is important for senior managers or others involved in the process to ask probing questions. Here are a few ways the team's reasoning should be explored:

1. In order to draw out more information from the design team, it helps to ask a lot of "why" questions, such as, "Why was a particular feature added?" or, "Why did you make this shape (or color, texture, etc.)?"

   The answers should have a reason grounded in good design. Examples of good questions and answers are:

   - *Why do you have the controls on the top, inside of the dishwasher door?* Because when they are on the outside of the door, they wear off over time when cleaned or frequently bumped.
   - *Why are the controls on top of the ovens?* So that children cannot easily reach them.
   - *Why did you choose this particular material?* Because it is made of sustainable fibers, and you can recycle the used item.

   You can start to see the types of questions you want to ask when assessing a new product design. Answers such as, "It just seemed to go with the rest of it," or, "I thought it looked good," are insufficient unless you happen to be designing jewelry for a client who just wants style.

   The answers can be a wealth of "free" copy for the marketing team. As the design team provides explanations for the product's benefits or superiority, the marketing team can take notes and use the material to inform content.

2. Ask the team if they have identified any concerns related to consumer use. This involves a balance between the team's instincts and the research/feedback the team received. The designer must have empathy for the consumer or user, but at some point, the designer has to decide on the direction, knowing that sometimes a customer cannot say why they are feeling a certain way, whether uneasy or elated. Will the team listen to one customer's convincing statement, or pull back to look at the wider set of users? Even experienced design teams grapple with this issue. There is no set answer, although a good rule of thumb would be to scrutinize the customer feedback for insights and inspiration that could help the greater target audience, if appropriate.

3. Ask the designer or team how the product or service was tested. Senior managers are not usually the target customer. Most inventions and designs target a wider age group or culture than what you, your team, and senior managers represent. If it is a toy, you should test it with children. If it is for a teenager, you should test it with teenagers. The design team cannot possibly know what will appeal to other age groups unless the specific age group is included in the process. When was the last time you bought a gift for someone only to realize that you bought it because *you* liked it, not because *they* did?

4. If possible, ask all of those involved on the team about their role and their thoughts about the project. Having a robust team of evaluators but a small group of decision makers can fizzle the excitement for the project. They may defer to the team leader, but at least all the team members will be acknowledged and addressed. This will help with team inclusiveness for the next project.

5. I run research projects that I call "expert design audits" for tech companies when they make a big change to their software programs. Expert audits can ease the mind of senior management when a tough or costly decision needs to be made. I pull experts from around the country (or world), have them sign nondisclosure agreements, and then evaluate the project. The expert evaluators point out problems and help prioritize the problems for senior management. If three of the expert evaluators point out the same issue, it's clearly a red flag that needs to be addressed.

What Does This Mean for the Future of Design?

- In the future, search programs will be able to quickly find and sort pertinent information. This could help companies and designers by reducing time spent gathering data.
- Improved virtual reality and augmented reality technologies can allow for better user testing in some cases. Virtual environments might help test expensive settings, such as theme parks or luxury experiences. Realistic simulation will help designers acquire more sophisticated and inspired feedback.
- New visualizing techniques will take you to a new level in design. Different visualization technologies will occur in the solution-generation stage and the final testing stages. Generative software will give designers more solutions than they could conceive on their own—and in a shorter amount of time. Objects may become more realistic as they are rendered and produced by robust 3-D printing. Animated imagery may enable space and experience designers to test more closely for emotions connected to products and services.
- Evaluating products with the correct target audience will become easier because you will be able to find real test subjects as opposed to convenient ones. You may have access to a wider database of consumers and users who can test products and services in a more efficient and representative way.

In the next chapter you will see what happens after the three stages of design. I will explore ways you can get your product into the global market and what you might expect in the future for marketing and distribution.

CHAPTER 8

Growing a Global Product

This chapter is for companies or people who have not yet "gone global" with their product but plan to, or for those who want to refresh their processes related to selling in foreign markets. Chapter 1 indicated which attributes can make a global product successful on the market, and chapter 2 discussed problems or opportunities that might affect international sales in the foreseeable future. Whether you are a one-person company or part of a large multinational company, certain practices should be followed to reduce risk when considering introducing a global product.

In this chapter, I will recommend activities that affect the design and innovation process, many of which involve extensive data collection in order to understand the history of the targeted area and the trending issues. Some of the recommended activities in this chapter may be dismissed as minor by larger companies, and some of the suggestions may seem insurmountable to the lone entrepreneur. But simply doing some of these recommendations will provide an informed foundation for the best decision-making.

Dorcy Inc. is a global distributor of portable lighting, such as flashlights, lanterns, and other handheld lighting devices. Life+Gear, owned by Dorcy, was founded after Hurricane Katrina in 2005, and manufactures personal lifesaving products.[1] Both companies have built their global sales through a combination of agents or distributors as well as direct sales to major retailers. I asked Tom Beckett, president/CEO of Dorcy, how an entrepreneur or small company could start to sell to other countries from a ground level. He said:

> "If you are targeting a country or region to sell to, you need to get on a plane and go there for at least a week or two and spend time exploring. Visit stores, talk to shop owners, check displays, ask about pricing, document the depth of assorted products. You will see how differently products are marketed in different countries. You will learn how some countries and regions welcome certain product features whereas another will not pay for certain features.
>
> If you've done some preparation, you may want to begin asking questions in relation to finding a good business partner. Ask who their distributor is, who stocks their products, and ask them if they could sell your product. Be prepared for some dead ends and blind alleys, as that is part of the process. To do additional preparation, you can go to the embassy or a trade organization for the country you are interested in and ask for names of reputable distributors for your type of product in that country and visit those distributors while you are there.
>
> Your ultimate goal is to find a good partner 'in country' who you can build a relationship with. If you are just starting out and have little or no brand equity in the markets you seek, this is an essential part of a long-term strategy. You must have someone with intimate knowledge of the market to get off the ground. Even if your product is universally adored by all, just navigating regulatory and logistics requires someone on the ground with firsthand knowledge. Do not be afraid to sacrifice some profit on the front end to get well-established. It will pay dividends many times over in the long run."[2]

There is no substitute for talking to the people and physically exploring the country or region in which you are interested. Documenting the trip is important as well; record towns, names of stores, and people you spoke with, so you can recall what you have learned, especially if you are sharing the information with a team at home. If you take photos in stores, you might want to first ask for permission. Are most of the products stocked in boxes and shelved, or are they hanging on a pegboard? Are they featured on endcaps, or buried in the back of the store? All of these details can help you make decisions when you return home.

In a *Harvard Business Review* article, "Finding the Platform in Your

Product," Andrei Hagiu and Elizabeth J. Altman list four strategies that can help you find customers and partners:

1. Partnering with third parties will open new venues and avenues for marketing and selling your products.
2. Connecting your customers to each other can be beneficial to all if they are not competing and are coordinating or cooperating on a product or service.
3. Connecting customer bases from two different products can bring new customer interaction or sales.
4. Sell to your customers' customers.[3]

Thinking ahead in regard to who your customers' customers are and what else they may like can help you position your product or service in a way to enhance this relationship without endangering your original relationship. You can start to add language around marketing and benefits related to another group of customers at the beginning of your design and marketing processes, provided that your original message does not get lost.

First: Data Gathering

In the next sections, I discuss data gathering pertinent to the product or service. Data gathering for the design process can come from online sources, experts, user research, networks, and traditional print sources, each with their own benefit. As data sets grow in value to users, you will probably have more opportunities to buy data sets and subsets of information.

Easy (and Consistent) Software Alerts

Today, just setting up systematic software alerts can allow you to build a lasting database for the future. Certain software packages will allow you to set up alerts for categories you are interested in or that are pertinent to your global product. Although there are several software alert systems on the market, Google makes an inexpensive but effective alert software. For countries that don't use Google, there are equivalent counterparts, though they may cost more or be less efficient.

The alert option is built into Google, and it sends emails to you when it finds a new result relating to your alert. It can search web pages, newspapers, blogs, and other content on the internet. Most importantly, you have the ability to refine your search by narrowing your search terms. Google also makes it easy to store and retrieve files.

The Ten Impact Factors from chapter 2 are a good starting point for setting up the depth and breadth of alerts in the country or region you are targeting. For example, if you want to know more information about Columbia, you can set up alerts related to climate, geography, history, culture, education, economy, technology, new products, new services, government, and so on. Having this information on file and available to the team can help in picking up trends and points of information to explore.

You may want to sell a product in a country with an upcoming election that will determine how welcome your new product or service will be. You will want to track that information, along with any other information related to it, such as cultural shifts, history, tariffs, currency, and labor rates.

Your company may be able to put together a repository of files to which colleagues can contribute. Someone in another division may be an undiscovered expert on a particular country; he or she may have a wealth of information, or a large photo collection. This data, when gathered, vetted, and organized, may be valuable for your company—and to other companies that might want to buy information and data on the same topic.

Currency Alert Checks

Currency fluctuations are another important item that can affect your project in a foreign country. Depending on the size of your project and your contracting needs, even a small fluctuation can affect profit margins. Rising labor rates, fluctuating currencies, and cost of materials can make a once profitable product or service too much of a risk.

User Research

In the early stages of the design process it is important to talk to potential customers and users. Getting information directly from users will provide insights that might affect the problem statement. For example, a company may believe a product needs to be completely redesigned, when in fact it might need to be simply resized.

When IKEA entered Asian markets, the company spoke with consumers about their lifestyles and living spaces. This data made it clear to IKEA that there was an issue with how their products fit within Asian apartments, which tended to be smaller in major cities due to the high cost of rent and ownership. Based on the findings, IKEA readjusted the size of some products to fit the living spaces in Asia. As IKEA expands in China and prepares to break into India, the company "trusts that its core concept, influenced by painstakingly acquired local knowledge, will, over time, give it an edge."[4]

When Roger Ball, a professor and associate chair at Georgia Tech School of Industrial Design, was designing sporting goods helmets in Asia, he noticed the fit was not right. It led to headaches and other discomfort for the Asians. This prompted him to conduct early research throughout China on the physical differences of head sizes between the East and West. He determined that the slight difference in head shapes (rounder in Asia and narrower in the West) was enough to necessitate a change in the design of helmets and headwear sold in Asia.[5]

Personal Networks

Valuable information about global regions can be found in many places, starting with your personal networks. Seek out as many people as possible who might know something that is pertinent to your project and can save you time, resources, and frustration. Friends (or coworkers) may have recently traveled to the region and have insights, updates, or impressions that may be valuable. You will not always get solid, objective research or data, but you will get anecdotal information that could spark an idea for marketing your product or service. You cannot assume, however, that someone in your country who is a foreigner will be knowledgeable about their home country. They may have lost touch with the rapidly changing culture in their former homeland, or they may not be interested in discussing it. Their observations may be no more or less subjective than your own or those of your team.

Professional Groups

Even if you do have a wide personal network or work for a large company with a powerful data-gathering capacity, you should join online groups and local groups that focus on global product development. Professional organizations can provide a wealth of information (and lifelong friendships). Professional organizations have always been a great resource for potential hires, salary information, best practices, and competitors' products. The difficulty, though, is keeping your idea under wraps while gathering information. After all, you cannot ask everyone in an informal conversation to sign a nondisclosure agreement.

At a past World Design Organization (formerly ICSID) meeting in Saudi Arabia, I learned that some products being sold in Europe were subject to lower tariffs because they had been packaged in certain parts of the Middle East. This information might not have come my way had I not been involved in that professional organization. Products did not have to be manufactured in the Middle East, they just needed to be packaged there. I would not have thought to go through a Middle Eastern country to sell to Europe (I was in the United States), but I passed this information on to several western companies that were interested in this type of distribution. Some people think the Middle East is unstable and difficult to penetrate when, in fact, most areas are politically stable and accommodating for business.

Economic and Trade Development Councils (Local and Global)

Trade development councils and regional economic councils can provide an abundance of accurate and up-to-date information on countries and regions. For general economic information about regions and countries, the website of the World Economic Forum (weforum.org) can be helpful. It also has world trends for commerce and best practices for selling products or services in other parts of the world.[6]

The Hong Kong Trade Development Council (HKTDC) started in 1966 and has 50 offices around the world and 13 on the Chinese mainland. The organization's website has information about conducting business worldwide, not just in Asia. The council also is a major information center for sourcing parts and products. The council's goal is to help businesses relocate to Hong Kong and use the manufacturers in the Pearl River Delta region, just over the border from Hong Kong.[7]

Most countries have a trade development group that you can approach for business information. These groups can provide valuable information on business practices and cultural issues. The consulates or embassies of most countries will be able to point you to reputable business trade groups within the country. If a city is large enough, such as New Delhi, Mexico City, or Shanghai, it will probably have its own business development group representing the city.

Hit the Stacks

Learning the history of the country and region is important. It is easy to believe you can find all the information you need online, and you don't need to go back to older information that might not be available online. Search secondhand bookshops or libraries for out-of-print books about the country or region. They may hold some clues for products and services that you can share with the team.

Second: Security and Your Team

Keeping product ideas and services confidential when you are actively searching for information is important. You must not reveal too much about your proposed product or service, and you want to keep the core aspects of your ideas confidential. Even searching online can give away information; ideas for new company names can be captured and recorded by software programs today that track which sites we visit online or which URLs we might want to purchase.

Filing provisional patents, which protects a product for approximately a year in the United States, can unfortunately "get the word out" before you are ready to offer your product or service to the marketplace. Searching for names for new websites (URLs) can put your idea out there before you are ready to commit to a new name. Keep things under wraps as much as possible until you are ready for the next stage.

Security is important when hiring new people. You may be interested in their coding abilities or some other talent, but be sure to hire for security as well. Screening the people you work with on a regular basis can be paramount to the security of your service or product.

Rick Cott, global head of Quantitative Qualitative Intercept (QQI), manages teams and organizations for ultra-high net worth families, both domestically and internationally. Cott says:

> "Much has been written about the need for cybersecurity from corporate espionage and other foreign powers. This is especially true when thinking about international business and the potential for theft of data. Any security expert will tell you that the safest way to protect your data is to first build teams of individuals who are trustworthy, understand the value of the company, and consider themselves members of the corporate community, in addition to having secure networks."[8]

The people whom you choose to hire and train are often a first line of defense in keeping your company's products, ideas, and data secure.

Third: Test Early, Test Often

Testing your product or idea does not have to be expensive. You can conduct informal interviews, construct low-resolution prototypes (foam or board models), or talk through scenarios, amongst many other methods to get a feel for whether your idea is on track. However, as you know, testing in the later stages with full-blown mock-ups, coded interactive programs, or full-scale environments requires extended time, money, and expertise.

Check in with your trusted network. An informal review or update of the project with colleagues and friends could be helpful throughout the project. You might not get detailed information, but you could get a "reality check" to identify possible flaws and determine whether you are on course in the design process.

Since it is expensive to do high-end testing, this informal feedback can yield critical information that helps in decision-making up until the manufacturing and final release of the product or service. Testing should be documented or captured for legal reasons as well. The best documentation is with video and written transcripts, which can be used to defend a patent. It can show details of the product design process that might not be obvious or have been forgotten, especially if several years have passed since the creation of the product. Documentation of testing also can be useful in case of wrongful-use lawsuits, and it can provide information that can be rolled into the next project or future upgrades.

Even if you perform extensive testing, and something does not seem correct or true, trust your instincts about what works and what does not. You may need to make difficult choices. It is easy to get confused or start doubting your direction when receiving test results and opinions, especially when they are conflicting. Keep your alerts and data coming. It is easy to stop paying attention to alerts when you are in the midst of a project. A key piece of information may show up in your alerts, and it could confirm that you are on the right track or indicate that you will have another obstacle.

The Next Manufacturing Center Opportunities

In an article about "The World's Next Great Manufacturing Center," author Irene Yuan Sun says, "Africa is in the early stages of a population boom that will reach 2 billion people by 2050, creating the largest pool of labor in the world."[9]

National governments in Africa are working to make their countries and regions attractive to businesses. Sun notes the "trodden path" of those countries and regions such as Japan, Korea, and China, which started out doing low-level manufacturing (e.g., ceramics or clothing) and gradually elevated to more sophisticated manufacturing, including electronic appliances and computers.[10]

Knowing where manufacturing occurs can benefit both start-ups and established global companies. Cost of labor is an obvious benefit. It also means that the company manufacturing in those regions will learn more about those cultures and signal opportunities for others. Provided the working conditions are fair and safe, new manufacturers can greatly assist the residents of a region, leading to better living conditions and wages, enabling people to escape poverty.

Hiring a Designer

If you have not hired a designer yet, know that designers can have specialties or propensities toward different areas in the arts or sciences. You can hire a designer who is an artistic craftsperson more interested in self-expression through a product. Another designer might be team-oriented and enjoy working on complex, scientific products. Of course, this is an oversimplification of the design field. But there are many types of designers on a creative continuum from arts to sciences, with some who are able to do both arts and sciences. Some designers prefer working on specific products, while others are generalists. Which designer you hire depends on what you need. The time spent learning about a designer's strengths and preferences during hiring will pay off down the road. To improve your odds, it helps to ask specific questions of the job candidates to learn their preferences. Hiring the wrong designer for a product or service can cost a company time, money, and other resources.

I once consulted with a technical company that had fired five designers in a row because none of them seemed to align with the work the company was doing. I realized that the company had been hiring designers who were more artistically than technologically inclined. These designers were from a good design school and had no interest in the technological side of the product. This resulted in a huge loss of time, both in hiring and trying to come up with a satisfactory design. In addition, money and resources were wasted on improbable and unacceptable solutions.

Suppose you are creating a housewares product and need work that is visual and expressive in nature. You can check the designer's portfolio for talent in this area. However, if your technical company manufactures appliances under a brand name, you may want to hire a designer who is interested in systematic and user interface work.

Most human resources departments do not have the in-house expertise to hire a designer, so search firms are often the answer. Such firms understand the field and the potential differences in education for a designer that comes from an art school or a university. They understand that differences of design education are important, but they can tune their search to fit your needs.

Schools are recognizing and addressing this distinction. Art schools and universities are blending their curriculum; art schools are integrating design research, and universities are reshaping the curriculum with more of an artistic aspect.

RitaSue Siegel, partner, RitaSue Siegel, tells us:

"Every company needs designers now more than ever. Experience design, interface design—the ability to focus on the user or customer is critically important to most companies, and most of the designers they now have started in traditional design silos. Many designers who are interested in upper management have moved into these strategic positions. They help structure the company in the C-level universe where plans are made, educating as they go about what's possible. There are designers with great ideas out there for the big picture of all of this, and they can help most any company plan for the future."[11]

Siegel is correct that many of the older designers who are now in corporations may have started during the "silo stage," where different areas of the company did not mix. However, these designers may have been waiting for more cross-disciplinary work, and they may be ready for next steps in their career, such as project management.

Angela Yeh, founder and chief talent strategist at Yeh IDeology, recruits designers for firms and advises on career paths. She says:

"Here is something that I want to completely debunk: that everybody is a designer. I have everybody from VPs to young students say, 'I'm a designer, and I solve problems.' Now if hundreds and thousands of people are saying the exact same thing but not clarifying their style of innovation, how is a business, an employer, a recruiter—how are these businesses to understand what design is? It's getting really hard for businesses to understand what each designer brings to the table."[12]

Hiring a design recruiter, especially if you are not familiar with the field, can save your company and design teams from a lot of unnecessary trial and error.

Additional Talents of Designers

Designers are trained to have a sensitivity to:

■ Different cultures
Designers look for clues in cultural differences that will help to design a better product. This means they can look for differences that might

enhance the visual aspects (form, color, texture), or the way a customer or user is approached, or any feature that might enhance the user experience.

- User experience
  Designers take time to break down actions into sequences in order to study what is happening when someone uses a product. In this systematic study of thoughts and actions, the designer can determine where something might have gone wrong, what was confusing, or what was not clearly delineated.

- Ethics
  Most designers adhere to ethical standards and will speak up if they think a product or practice is unsafe. They will point out materials that may be problematic for people, pets, or the environment. They will also make certain that user testing does not harm anyone, and ensure that ethical guidelines are followed for research.

- Sorting out the best solutions
  Designers have the ability to look at information and come up with options that might satisfy the problem statement. They can think visually and imagine products and services with different attributes. This ability enables them to mentally rule out solutions that won't work and focus on the ones that will.

- Visualizing ideas that do not exist
  It can be very difficult to visualize a product. One of the best attributes of designers is their ability to create an image of the product (or service) so that others can visually experience it. Not all team members are visually oriented, so it can be difficult for them to envision an idea. So it helps when the idea is visually represented. These visualizations also help the team reflect on the product or service in a way that is different from discussing it or seeing it in a realistic state. When the team sees ideas in visual form, it is often easier to determine what is wrong.

- Choosing sustainable materials and manufacturing methods
  Designers are educated today to automatically choose materials that are not toxic to humans, animals, or the environment. They will also work with engineers to find the best possible manufacturing methods, even if it means changing the design to accommodate a different material or a better process.

- Working alone or in a team

Designers may find it more stimulating to work in a team environment. The team's energy often leads to more ideas—and move evolved ideas—than those produced by someone working alone. However, some designers prefer to take in as much information as possible by themselves, and then bring some solutions to the team to build upon.

What Does This Mean for Future Global Product Design, Manufacturing, and Distribution?

More opportunities than ever for timely information and data gathering
Data and information are easier than ever to find and use to make decisions. This benefit will likely increase in the future, as data gathering and organizing will become more easily accessible and refined.

Hiring the right designer to fit with your company
If you have a highly technical company and you need a designer to work on a product that is systematic with frequent technology upgrades, you do not want to hire an artistic designer. Alternatively, if you are developing products that require attractive and unique forms, colors, shapes, or textures to enhance an environment, you may not get wildly unique and beautiful pieces from a designer who specializes in information design. Some designers may be excellent at everything, but it is very difficult to find them, and you shouldn't bank on it. Also, some companies may feel they need "young" designers for their products or services, but most designers get better with experience, as they have accumulated knowledge and expertise that can save the company money and frustration.

Starting out with a secure team
You have the option to hand pick and screen your team members for loyalty, security, and confidentiality. Screening employees is an ongoing process, not just a one-time event. Over time you need to remind your company and team members to adhere to strict security practices at all times.

Start by providing value to a region
Providing value to a region may help your company or product weather the political landscape, such as protectionist trends. When you start your

venture by creating value to a country or region, it will have a better chance of surviving political turmoil.

The opportunity and complexity of finding global partners will increase
As more areas in the world open to business, become easily accessible, and provide infrastructure, your opportunity to find global partners and new platforms will increase. However, with this increase in opportunity comes an increase in complexity. It may be best for a young company to do a quick consideration of the best areas of the world in which to do business, but then focus on one.

CHAPTER 9

Summaries and Conclusions: Design Reasoning with Insights

What Will Make a Global Product Successful?

In the past, sales and aesthetics were enough to consider a product successful. But today, consumers want additional benefits, such as a company mission to help the world and their community. Nowadays, successful brands are built on good company deeds over time.

Impacts on Product Success or Failure

The increased complexity of designing a product or service for global use has made data gathering a crucial step in the design process. More information is needed for decision-making to meet the rapid changes in governments and cultures worldwide. One simple way to capture these changes is through an alert system, such as Google's system, which provides geographically targeted and timely information about cultural trends, currency and labor market fluctuations, geography and climate issues, government edicts, and anything else that could impact your decisions. In the future we will have several options for mitigating risk coming from big data mining.

The Design Process, Design Thinking, and Innovation

The design process has an overarching methodology that begins with a problem (or opportunity) statement and ends when the chosen design is evaluated. This process is often represented as a linear process, but in practice the steps are highly iterative. Design thinking can occur throughout the design process, starting with a divergent, open type of thinking in order to generate as many solutions as possible. Insights occur from connecting and synthesizing the data with a skill called design reasoning. Design reasoning may take place throughout the process as a way to think about potential solutions. As a solution is chosen, refined, and tested, the design thinking becomes convergent, narrowing in on final solutions.

Attributes of Product Design

Early designs focused on utility or beauty (or both, depending on the product). These attributes are still highly desirable, but expectations from consumers have increased the need for new attributes, such as sustainable materials and manufacturing processes, appropriate ethics and technologies, and a social element that contributes to the community or country where it is made.

Designers in the future may work mostly on the experience a design product or service provides. They may also design interactions for artificial intelligence systems, robot-to-robot, robot-to-human, and human-to-virtual spaces.

How to Make a Successful Environment for Innovation

Designing a great creative space in which to work is not just about looking good; creative space is a highly visible commitment by the company to support design and innovation. Creative spaces are useful for supporting the work of the design team. Walls with notes to capture ideas, room to hang images for critique, comfortable seating, digital whiteboards, video recording, and more can serve as the operations space for the team. The ability to spread out ideas in front of the team—and revisit those ideas as needed—is highly conducive to creative problem solving.

In the future, however, you may need to rethink the spaces used in the design process when generative software becomes more refined and available. In addition, since products and services will become more complex, you may have to accommodate larger teams with more diverse expertise. Virtual work spaces may be needed as well. There may be a need for design moderators who keep the process on track and introduce other key team members as needed.

Managing a Global Design Team

Managers of global teams will be pleased by future improvements in video conferencing, scheduling assistance, and video capturing. Global design managers will still need to deal with cultural differences. In the future, team and project management software will be further improved to assist with some of the tasks, such as manufacturing schedules, billing, gathering data, and other daily tasks you may hand off to a digital assistant.

How to Evaluate Product Concepts

Evaluating products and services in the past was simple, if not always satisfactory. Either the boss or the marketing director would make the final decision. Today, user research is vital to help understand what product or service is most appealing to a target customer. User research will increase throughout the design process in the future in order to learn about what is successful and help the design team to mitigate risk of an inappropriate product design. User research methodology will become more adaptable to different cultures as global sales increase. We may also see an increase in reliable databases that can make decision-making much easier.

Technologies for testing products and services in the future may become more varied as sensor technologies of all kinds improve. Facial recognition along with emotion-detecting software may provide improved testing of products and services as a regular exercise and be carried out remotely, as well.

Growing a Global Product

Both new and established companies have more opportunities to find platforms for global sales because of the internet. New technologies will make it easier to find critical information, engage new partners, and research their credibility.

With new manufacturing centers emerging in regions around the world, there are more opportunities to sell products related to that manufacturing center. Infrastructure for technology and distribution is important, as is learning whether there is a demand for your product or service in a particular culture.

Future Issues for Everyone

The future of design will involve everyone, but, where you will see the most concern is the issue of technology taking over various roles of human life. If technology is considered appropriate, useful, and for a good cause, you will most likely find it acceptable, but forced, unwanted behaviors through unwanted technology will be met with resistance.

Technologists vs. New Luddites

In coming years, the two major groups that will impact design (and lifestyles) are the technologists (people who are enthusiastic about the role of technology in the future) and the new Luddites (people who value the natural human body and mind, without alteration, and want to slow down the general encroachment of technology).

The technologists will be excited about what new technologies will be able to do. They believe that technology will help eradicate diseases and reduce human suffering. From their point of view, new technologies will:

- Allow exploration of the space within and around humanity, and help people to learn new ways of being, living, working, and playing.
- Help remedy some of the mistakes of the past, such as policies that harm the environment.
- Solve world hunger.

- Help bring millions of people out of poverty and provide education around the world.
- Revitalize cities and protect rural areas.
- Make people the best they can be.

However, technologists may pursue applications that could be considered unethical and dangerous to the human species.

In the future you will hear protests from the new Luddites who may argue for less technology. There are people who dislike technology and believe it is already integrated too deeply into daily life. They believe they are not given choices when it comes to new technologies, and they believe technology will be the destruction of us all due to tampering with the environment, the human body, or warfare.

Of course, represented above are two conflicting extremes around the technological issues of tomorrow. These extremes are labeled in a very simplistic way here, but the issues remain complex.

Ethics in Design and Technology

As designers and design teams work with more technologies, they may encounter questionable practices that begin to seem normal. Did game designers or social media designers realize their products were addictive for some people? If they did, that is an ethical problem. If they just wanted to make a fun game or social media platform, then they could be considered naïve.

When research proved that cigarettes cause cancer,[1] some advertising designers and marketers chose to no longer work with the tobacco industry. This meant finding a new job or project because their personal ethics made it difficult for them to promote smoking. Will the designers of today and tomorrow apply their personal ethics to stop inappropriate messages and harmful practices?

Meeting in the Middle

Technologist and humanistic agendas may clash far into the future. How much technology do we accept into our lives, and when is a smart machine too smart? Mediation may be needed to create the best of both worlds, and design thinking and reasoning may play a prominent role in this regard, by using the design process to sort out issues with users and giving a voice to all.

Design teams can work on "big picture" problems for the common good. If companies would dedicate some of their resources and pull teams together to consider the pros and cons of extreme outcomes, they may be able to avoid some of the negative aspects of technology and instead come up with honest ways to build trust in users.

I am not advocating that the clash between the technologists and new Luddites becomes a design problem, but I am saying that the design process, design thinking, and design reasoning can lead to potential solutions, or at least help to further define the problem statement. It is a very large conversation and many voices are needed for appropriate solutions.

Design Reasoning Research

If you examine the design process closely, you find an embedded process called design reasoning. Design reasoning helps designers and design teams to synthesize large amounts of information and stimuli, and look for insights and unique connections in the data and stimuli.

Design reasoning gets to the heart of the matter of creating new ideas and innovations. The designer finds herself considering information in many forms in relation to the problem and the solution. The act of design reasoning can occur through all phases of the design process.

Design reasoning at a glance can seem more chaotic than systematic. The chaos is a result of trying to bring all of the information together to find connections that could result in solutions. If you listened to someone using design reasoning, you might not understand how they could possibly narrow down anything; their attention sometimes jumps from one point to another in a nonlinear way. A simplified example of stream of consciousness chaotic design reasoning might be:

> "I need to make inflatable rafts for disaster areas. I think a yellow reflective material would be good, as it is highly visible in daylight and at night. What kind of material should it be? I saw a tough nylon fabric over inflated pontoons the other day, maybe we can find and use that

if it stands up to the testing. The experience for these people in the raft has to help them feel safe and relax, if possible. Should there be a soothing pattern on the inside and have some type of embedded lighting? Let me look at those new floating evacuation boats for disabled people who are caught in floods. I will draw some of these ideas and patterns so I remember them. The flood survivors I spoke with yesterday were amazing. The way they had to survive a flash flood was terrifying. Oh, here is data that has just come in for ergonomic safety seating in rescue craft" . . . and so on until solutions begin to present themselves.

A more systematic version of design reasoning is when inspiration or ideas take hold and the designer or design team go down a mental path to explore its viability. Other stimuli and data are excluded in order to further focus on one idea. An example of this thinking might be: "I have a new configuration of the flood raft. It can hold people, pets, and it has a few small areas of protected dry storage for small items such as documents, cards, or keys. I have made the interior a soothing green color, with embedded lighting on the outside for visibility to rescuers, and on the inside so the occupants can see each other. I have added a system that can slow the raft down in rushing water by adding these side panels. I have included connectors to hook life rafts together in case there is a larger number of people to rescue. When I get the full-size model, I should set up meetings with the disaster relief folks to see if this product can be improved or modified for use in other disasters. I should also get feedback from some of the people who were saved in a life raft" . . . and so on until the concept is tested.

Design reasoning is worth studying and analyzing for the future because it provides clues to how new ideas emerge and innovations come about. This type of reasoning occurs within a larger framework of the design process with design thinking.

The Future for Designers

I hope that the best of design skills will not be lost in our transition to a new design era. I believe we will still need to design beautiful, intriguing, useful, and tactile things, maybe not in the amounts we have had before, because of sustainability and the taxed ecosystem of our planet. However, designers of the future will need to greatly broaden their education and research skills if they are to design for the new technologies that will be embedded in our spaces, products, and experiences. Designing for all the senses (in real as well as virtual environments) will become more important for future solutions. Designers may also be designing for product-to-product experiences as well as product-to-human experiences. Designers may need training to learn to work with complex teams, adapt to new forms of design research, and learn new software and technologies. We may need more design degrees that allow a focus or specialty, such as a Design PhD or Doctorate in Design.

Complexity will come from shifting cultures and governments, speedier processes, and trying to plan for change in an ever-changing world. Design thinking, design reasoning, and visualization and communication techniques will likely be adopted, learned, and used by other disciplines. Designers may find themselves working with team members from different industry backgrounds who are aware of design processes and methodologies, enabling the team to move at a fast pace. Ideally, companies in the future will have the design processes embedded in their day-to-day operations.

Communication and data-gathering technologies may make decision-making easier in some cases because the amount of data gathered can inform a decision to lessen risk.

I end with a quote from Suzanne LaBarre, editor of Co.Design.[2] She says:

> "We hope to convey the message that design is *the* story of business in the twenty-first century. You can't talk about business without talking about design. . . . We hope to tell deeper stories about the stunningly complex problems facing companies and organizations, whether it's how to scale, how to develop ethical AI, or how to redesign an entire generation."[3]

These are no small tasks, and LaBarre is right. In light of this, I recommend that companies and organizations make no small plans in relation to design. The future is design, and the future needs your attention and engagement.

Endnotes

Introduction

1. Hanson Robotics. "Sophia." hansonrobotics.com.
2. Ibid.
3. Heather McGowan (Cofounder Work to Learn), interviewed by Lorraine Justice, April 2018.
4. World Economic Forum. "Eight Futures of Work: Scenarios and Their Implications," January, 2018. weforum.org/whitepapers/eight-futures -of-work-scenarios-and-their-implications.
5. Ibid.
6. Winick, Erin. "Every study we could find on what automation will do to jobs, in one chart," *MIT Technology Review*, January 25, 2018. technologyreview.com/s/610005/every-study-we-could-find-on-what-automation-will-do-to-jobs-in-one-chart.
7. Ibid.
8. Yu, Howard. "The hyper vision of almost every disruptive technology," South China Morning Post, February 2, 2018. scmp.com/business/article /2131627/hyper-vision-almost-every-disruptive-technology.
9. DLD. "Digital Life Design." dld-conference.com.
10. Christopher Keller (Head of Group Business Development of the London Telegraph), interviewed by Lorraine Justice, January 2018.
11. Ibid.
12. Martin Wezowski (Chief Designer and Futurist of SAP), interviewed by Lorraine Justice, January 2018.
13. Ibid.

14. Pelegrin, Williams. "What do people actually use Amazon Echo and Google Home for?" Android Authority blog, July 12, 2017. android-authority.com/amazon-echo-google-home-ifttt-786753/.

15. Ibid.

16. Kerr, Breena. "The New Generation of Virtual Personal Assistants." *New York Times*, March 25, 2018.

17. Sorkin, Andrew Ross. "A 'Gadget Junkie,' Wearing His Tech and Covering Deals," *New York Times*, February 1, 2018.

18. Norman, Donald, email, February 2018.

19. Ibid.

20. Keen, Andrew. *How to Fix the Future: Staying Human in the Digital Age*. London: Atlantic Books, 2018. 51.

21. McGowan, interview.

22. Patricia Moore (President of MooreDesign Associates), interviewed by Lorraine Justice, June 2018.

23. Carey, Benedict. "Brain Implant Enhanced Memory, Raising Hope for Treatments, Scientists Say." *New York Times*, February 7, 2018, A21.

24. Jeremy Haefner (Provost and Executive Vice Chancellor, University of Denver), interviewed by Lorraine Justice, May 2018.

25. Ibid.

26. Social Robot Design Challenge. "What is the design challenge?" depts .washington.edu/designme/.

27. Linn, Allison. "XiaoIce, Microsoft's Xiaolce is an AI bot that can also converse like a human." CNET, May 22, 2018. https://www.cnet.com/news /microsofts-xiaoice-is-an-ai-bot-that-can-also-converse-like-a-human.

28. Tim Fletcher (President of One BusinessDesign), interviewed by Lorraine Justice, April 2018.

29. Tim Fletcher, "Tools and Methods 006 - Design Thinking in Paradise: Facilitating Change," One BusinessDesign blog, April 10, 2018. www .onebusinessdesign.com/blog/2018/4/10/tools-and-methods-006 -design-thinking-in-paradise-facilitating-change.

Chapter 1

1. Patagonia. "Patagonia's Mission Statement." www.patagonia.com /company-info.html.

2. Ikea. "Our vision and business idea." www.ikea.com/ms/en_SG/about _ikea/our_ business_idea/index.html.

3. Hekkert, Paul, and Pieter Desmet, "Framework of Product Experience." *International Journal of Design* 1, no. 1 (2007). 58.

4. Forbes. "The World's Most Valuable Brands." https://forbes.com /powerful-brands/list#tab:rank.

5. Ibid.

6. Justice, Lorraine. *China's Design Revolution*, p. 113. Massachusetts: MIT Press, 2012.

7. "China's urban middle class key to brand success as consumers move from price to premium in the 2017 BrandZ Top 100 most valuable brands ranking." BrandZ and Kantar Millward Brown Press Release. https:// www.wpp.com/-/media/project/wpp/files/news/kantar_pressrelease _brandz_china_mar17.pdf.

8. Chai, Catherine. "The Top Brands in Asia to Watch in 2018." Branding in Asia. January 16, 2018. https://brandinginasia.com/top-brands-asia -watch-2018/.

9. Africa Media Agency. "Brand Africa 100: Africa's Best Brands 2017/18." https://www.africa.com/brand-africa-100-africas -best-brands-2017-18/.

10. Tucker, Ross. "The 50 Most Valuable Latin American Brands of 2018." Kantar US Inights. https://us.kantar.com/business/brands/2018 /top-50-latin-american-brands-2018/.

11. The Statistics Portal. "Retail unit sales of rice cookers in the United States from 2010 to 2017." statista.com/statistics/514613 /us-retail-unit-sales-of-rice-cookers.

12. iF World Design. "iF: Design for Good." https://ifworlddesignguide .com/press-about/about-if/the-if-story.

13. Good Design Australia. "Design for a Better Australia." https://good -design.org/about/.

14. Ibid.

15. Good Design Australia. "Categories & Criteria." https://good-design .org/good-design-awards/categories-criteria.

16. International Designers Society of America. International Design Excellence Awards. www.idsa.org/IDEA.

17. Ibid.

18. Hong Kong Design Centre. "DFA Design for Asia Awards." https:// dfaa.dfaawards.com/.

19. Ibid.

20. Hong Kong Design Centre. "DFA Design for Asia Awards, About." https://dfaa.dfaawards.com/background/.

21. Design to Improve Life. "No More White Tea Cups!" https://designto improvelife.dk/history/.

22. Ibid.

23. Waughray, Dominic Kailash Nath. "Why the Global Goals are a golden opportunity for all of us." World Economic Forum, Sustainable Development. September 24, 2015. http://www.weforum.org/agenda /2015/09/why-the-global-goals-are-an-opportunity-for-all-of-us/.

24. World Economic Forum. "What are the Sustainable Development Goals?" https://www.weforum.org/agenda/2015/09/what-are-the-sustainable -development-goals/.

25. Ibid.

Chapter 2

1. The Economist. "The world's most valuable resource is no longer oil, but data." May 6, 2017. https://www.economist.com/leaders/2017/05/06 /the-worlds-most-valuable-resource-is-no-longer-oil-but-data.

2. Frankopan, Peter. *The Silk Roads: A New History of the World*. New York: Vintage Books, 2017. xvii.

3. Weiner, Eric. *The Geography of Genius: Lessons from the World's Most Creative Places*. New York: Simon & Schuster Paperbacks, 2016. 68.

4. Ibid.

5. Schaffhauser, Dian. "Universities in Pittsburgh and Paris to Host New AI R&D Centers." Campus Technology, May 1, 2018. https:// CampusTechnology.com/articles/2018/05/01/universities-in -pittsburgh-and-paris-to-host-new-ai-rd-centers.aspx.

6. Wee, Sui-Lee,and Paul Mozur. "Building the Hospital of the Future." *New York Times*, February 1, 2018.

7. Quarmyne, Nyani, and Kevin Granville. "Data Roaming, to the Extreme." *New York Times*, January 6, 2018.

8. Robles, Frances. "Thousands Still in the Dark as Some Power Workers Exit Puerto Rico," *New York Times*, February 27, 2018.

9. Williams, Lauren. "Smart drones can identify sharks near shore." *USA Today*, December 23, 2017.

10. Justice, Lorraine. *China's Design Revolution*. Massachusetts: MIT Press, 2012.

11. Jacobs, Andrew. "In War on Obesity, Chile Slays Tony the Tiger." *New York Times*, February 8, 2018.

12. Bradsher, Keith, and Noam Sheiber. "Growing Pains of Going Global." *New York Times*, November 9, 2018.

13. Manjoo, Farhad. "Digital Addiction Stirs Worry Even in Its Creators." *New York Times*, February 12, 2018.

14. Ibid.

15. Walsh, Declan. "In Egypt, Encryption is Essential." *New York Times*, November 9, 2017.

16. Keen, Andrew. *How to Fix the Future: Staying Human in the Digital Age*. London: Atlantic Books, 2018. 75.

17. Ibid.

18. Alderman, Liz, Elian Peltier, and Hwaida Saad. "What Price for Profit in Syria?" *New York Times*, March 11, 2018.

19. Moore, Jina. "Kenyan Officials Keep 4 TV Stations Dark for Days." *New York Times*, February 3, 2018.

20. "The importance of independent validation." Pictet Report, 21, 2017, 11.

Chapter 3

1. Bürdek, Bernhard E. *Design: History, Theory and Practice of Product Design*. Germany: Birkhäuser Verlag GmbH, 2015. 108.

2. Ibid.

3. Tucker Viemeister, interviewed by Lorraine Justice, March 2018.

4. Csikszentmihalyi, Mihaly. *Creativity: Flow and the Psychology of Discovery and Invention*. New York: HarperCollins, 1996. 58.

5. Ibid., 72.

6. The Conversation. "From the mundane to the divine, some of the best-designed products of all time." https://theconversation.com /from-the-mundane-to-the-divine-some-of-the-best-designed -products-of-all-time-72697.

7. Kelley, Tom, and Jonathan Littman. *The Art of Innovation*. New York: Doubleday, 2001. 31.

8. Fernandez, Luis Rajas. "Is Innovation a Buzzword?" Medium. https://medium.com/@LuisRajas/is-innovation-a-buzzword-83fccb069f58.

Chapter 4

1. Baron, Katie. "Loving the Alien: Why AI Will Be the Key To Unlocking Consumer Affection." Forbes. www.forbes.com/sites/katie-baron/2018/05/14/loving-the-alien-why-ai-will-be-the-key-to-unlocking-consumer-affection.

2. Cagan, Jonathan, and Craig M. Vogel. *Creating Breakthrough Products: Revealing the Secrets that Drive Global Innovation*. New Jersey: Pearson Education, 2013. 73.

3. Ibid.

4. Stack, Liam "H&M apologizes for 'Monkey Image' featuring a black child," *New York Times*, https://www.nytimes.com/2018/01/o8/business/hm-monkey.html.

5. Tom's Shoes. "Improving Lives." https://www.toms.com/improving-lives.

6. Maize. "How Jack Ma is Changing the Chinese Retail Game," June 22, 2018. https://www.maize.io/en/content/how-jack-ma-is-changing-the-chinese-retail-game.

7. Coach. "Signature." https://www.coach.com/shop/signature.

8. Myers, Jack. *The Future of Men: Masculinity in the Twenty-First Century*. California: Inkshares, 2016. 273.

9. Shedroff, Nathan. *Design is the Problem: The Future of Design Must be Sustainable*. New York: Rosenfeld Media, 2009. 204.

10. Kahane, Josiah. *The Form of Design: Deciphering the Language of Mass-Produced Objects*. Amsterdam: BIS Publishers, 2015. 45.

Chapter 5

1. Patton, Drew. "The new corporate campus." Workdesign Magazine, May 24, 2016. /https://workdesign.com/.

2. Cloudpeeps. "Top 10 companies winning at remote work culture and their secrets," September 8, 2015. //blog.cloudpeeps.com/top-10-companies -winning-at-remote-work-culture/.

3. Quinton, Laura. "The Future Smart City, Interview with Anne Stenros." GOinternational Finland, May 2, 2018. http://gointernational.fi /the-future-smart-city-anne-stenros/.

4. Ibid.

5. Alexander and Jochen Renz (Founders of New Mobility Consulting), interviewed by Lorraine Justice, May 2018.

6. Benedictus, Leo. "Chinese city opens 'phone lane' for texting pe-destrians." *Guardian*, September 15, 2014. https://www.theguardian .com/world/shortcuts/2014/sep/15/china-mobile-phone-lane -distracted-walking-pedestrians.

7. Gensler. "Design Forecast 2015: Top Trends Shaping Design." www .gensler.com/design-forecast-2015-the-future-of-workplace.

8. Kasriel, Stephane. "Cities are Killing the Future of Work (And the American Dream)." *Fast Company*, January 18, 2018.

9. Burdette, Kacy. "See Photos of Tech Companies' Futuristic Headquarters." *Fortune*, Oct. 25, 2017. http://fortune.com/2017/10/25 /see-photos-of-tech-companies-futuristic-headquarters/.

10. H-Farm. "About." www.h-farm.com/en/about.

11. Ibid.

12. 360° Steelcase Global Report. "Engagement and the Global Workplace." 8.

13. Ibid., 3.

Chapter 6

1. Kotter, John. *Leading Change: An Action Plan from the World's Foremost Expert on Business Leadership*. Boston: Harvard Business Press, 2012. 167.

2. Ibid.

3. Kent, Chuck. "Innovation, conversation and leadership: Nick Partridge from LPK." Lead the Conversation, January 8, 2018. https:// leadtheconversation.net/2018/01/08/innovation-conversation-and -leadership-nick-partridge-from-lpk/.

4. Alex Chunn (Vice President of TTI), interviewed by Lorraine Justice, July 2018.

5. Bariso, Justin. "Google spent years studying effective teams. Here's the 1 thing that mattered most." *Inc.*, August 6, 2018. https://www.inc .com/justin-bariso/google-spent-years-studying-successful-teams -heres-thing-that-mattered-most.html.

6. Ibid.

7. Haas, Martine, and Mark Mortensen. "The Secrets of Great Teamwork." *Harvard Business Review*, 71.

8. Ibid, 75.

9. Dziersk, Mark. "An Innovation Equation." LinkedIn. https://www .linkedin.com/pulse/innovation-equation-mark-dziersk/.

10. Sawyer, Keith. *Group Genius: The Creative Power of Collaboration*. New York: Basic Books, 2017. 34.

11. Meyer, Erin. "Being the Boss in Brussels, Boston, and Beijing." *Harvard Business Review* (July-August 2017). 72.

12. Ibid., 73.

13. Lubin, Gus. "24 Charts Of Leadership Styles Around The World." *Business Insider*, January 6, 2014. www.businessinsider.com/leadership -styles-around-the-world-2013-12.

14. Ibid.

15. Ibid.

16. Ibid.

17. Ibid.

18. Eppinger, Steven D., and Anil R. Chitkara. "The Practice of Going Global." *MIT Sloan Management Review*, Summer 2009.

19. Neeley, Tsedal. "Global teams that work." *Harvard Business Review*, October 2015

20. Qifeng Yan (CEO Loobot), interviewed by Lorraine Justice, May 2018.

21. Chunn, interview.

22. Leach, Whitney. "This is where people work the longest hours." *World Economic Forum*, January 16, 2018. weforum.org/agenda/2018/01 /the-countries-where-people-work-the-longest-hours/.

23. James Ludwig (Vice President of Steelcase), interviewed by Lorraine Justice, July 3, 2018.

24. Ibid.

25. Chunn, interview.

26. Brown, Bruce, and Scott D. Anthony. "How P&G Tripled Its Innovation Success Rate." *Harvard Business Review*, June 2011. 64–71.

27. Ibid., 67.

28. Dann, Jeremy B, Katherine Bennett, and Andrew Ogden. "Xiaomi: Designing an Ecosystem for the 'Internet of Things'." Case Study, Lloyd Greif Center for Entrepreneurial Studies, Marshall School of Business, University of Southern California. 2017. https://hbsp.harvard.edu /product/SCG527-PDF-ENG.

29. Girling, Rob. "AI and the future of design: What will the designer of 2025 look like?" O'Reilly. www.oreilly.com/ideas/ ai-and-the-future-of-design-what-will-the-designer-of-2025-look-like.

30. Ibid.

31. Dickson, Ben. "7 surprising companies where you can work on cutting-edge AI Technology." The Next Web. https://thenextweb.com /artificial-intelligence/2018/07/05/companies-work-ai-technology/.

Chapter 7

1. Brown, Tim. *Change by Design: How Design Thinking Transforms Organizations and Inspires Innovation*. New York: HarperCollins, 2009. 70.

2. Martin, Bella, and Bruce Hanington. *Universal Methods of Design*. Massachusetts: Rockport Publishers, 2012.

3. Rochester Institute of Technology. "Golisano Institute for Sustainability." www.rit.edu/gis/.

4. Ibid.

Chapter 8

1. Tom Beckett (President/CEO of Dorcy), interviewed by Lorraine Justice, July 2018.

2. Ibid.

3. Hagiu, Andrei, and Elizabeth J. Altman. "Finding the Platform in Your Product." *Harvard Business Review*, July-August, 2017. 95–100.

4. Ringstrom, Anna. "One size doesn't fit all: IKEA goes local for India and China." Reuters, *The Globe and Mail*, May 11, 2018. www .theglobeandmail.com/report-on-business/international-business

/european-business/one-size-doesnt-fit-all-ikea-goes-local-for-india -china/article9444097/.

5. Roger Ball (Professor at Georgia Tech), interviewed by Lorraine Justice, June, 2018.

6. World Economic Forum. "Our Mission." www.weforum.org/about/world-economic-forum.

7. Hong Kong Trade Development Council. "About HKTDC." http://aboutus.hktdc.com/en/#global-network.

8. Rick Cott (CEO of Quantitative Qualitative Intercept), interviewed by Lorraine Justice, February, 2018.

9. Yuan Sun, Irene. "The World's Next Great Manufacturing Center." *Harvard Business Review*, May-June 2017. 125.

10. Ibid.

11. RitaSue Siegel (President, RitaSue Siegel Resources), interviewed by Lorraine Justice, June 2018.

12. Angela Yeh (Founder and Chief Talent Strategist at Yeh IDeology), interviewed by Lorraine Justice, April 2018.

Chapter 9

1. American Lung Association. "Tobacco Industry Marketing." http://www.lung.org/stop-smoking/smoking-facts/tobacco-industry -marketing.html.

2. LaBarre, Suzanne. "Co.design joins fastcompany.com." *Fast Company*. fastcompany.com/90180557/co-design-joins-fastcompany-com.

3. Ibid.

The publisher has used its best efforts to ensure that any website addresses referred to in this book are correct and active at presstime. However, publisher and the author have no responsibility for the websites and can make no guarantee that a site will remain live or that the content will remain relevant, decent, or appropriate.

Lorraine's Work
with Clients

Lorraine Justice, PhD, is a globally-recognized expert in design. She helps her clients think about where they are on the continuum of technology and human behavior and design products and services for maximum success.

She works with leaders and teams to:

- Identify all relevant trends in product and service design that could impact success in a given market
- Understand how to build and manage a successful innovation team
- Ask the right questions about functionality so developers focus on the user and not the technology
- Get "from here to there," addressing feasibility and staging

To learn more about Dr. Justice's work with clients, visit LorraineJustice.com

Bring "The Future of Design" Presentation to Your Organization

We interact with hundreds of products and hundreds of different sensory stimuli every day, from the moment our bed gently shakes us awake and takes our vitals until the minute our head hits our temperature-controlled pillow and we close our eyes on the sun-setting clock we asked to wake us up at seven in the morning. This talk takes audiences on an informative and amusing journey of technology and lifestyle to sensitize your audience to ask important questions such as, "How much technology does your customer want?"

Audiences walk away understanding:

- The broad range of consumer technology likes and dislikes that are not just a generational issue
- An in-depth understanding of what your audience members would personally need to begin innovation with big questions
- Where we are as a society in the continuum between technology and the human condition

To invite Lorraine Justice to speak, visit LorraineJustice.com

Index